T0296257

CAMBRIDGE LIBRARY COLLECTION

Books of enduring scholarly value

Physical Sciences

From ancient times, humans have tried to understand the workings of the world around them. The roots of modern physical science go back to the very earliest mechanical devices such as levers and rollers, the mixing of paints and dyes, and the importance of the heavenly bodies in early religious observance and navigation. The physical sciences as we know them today began to emerge as independent academic subjects during the early modern period, in the work of Newton and other 'natural philosophers', and numerous sub-disciplines developed during the centuries that followed. This part of the Cambridge Library Collection is devoted to landmark publications in this area which will be of interest to historians of science concerned with individual scientists, particular discoveries, and advances in scientific method, or with the establishment and development of scientific institutions around the world.

An Elementary Treatise on Electricity

James Clerk Maxwell (1831–1879), first Cavendish Professor of Physics at Cambridge, made major contributions to many areas of theoretical physics and mathematics, not least his discoveries in the fields of electromagnetism and of the kinetic theory of gases, which have been regarded as laying the foundations of all modern physics. This work of 1881 was edited from Maxwell's notes by a colleague, William Garnett, and had formed the basis of his lectures. Several of the articles included in the present work were also included in his two-volume *Treatise on Electricity and Magnetism* (1873), also reissued in this series. The preface indicates that the two works were aimed at somewhat different audiences, the larger work assuming a greater knowledge of higher mathematics. Maxwell had also modified some of his methodology, and hoped to encourage the reader to develop an understanding of concepts relating to electricity.

An Elementary Treatise on Electricity

James Clerk Maxwell
Edited by William Garnett

CAMBRIDGE UNIVERSITY PRESS

Cambridge, New York, Melbourne, Madrid, Cape Town,
Singapore, São Paolo, Delhi, Tokyo, Mexico City

Published in the United States of America by Cambridge University Press, New York

www.cambridge.org
Information on this title: www.cambridge.org/9781108028783

This edition first published 1881
This digitally printed version 2011

ISBN 978-1-108-02878-3 Paperback

Clarendon Press Series

AN

ELEMENTARY TREATISE

ON

ELECTRICITY

MAXWELL

Clarendon Press Series

AN ELEMENTARY TREATISE ON ELECTRICITY

BY

JAMES CLERK MAXWELL, M.A.

LL.D. EDIN., D.C.L., F.R.SS. LONDON AND EDINBURGH
HONORARY FELLOW OF TRINITY COLLEGE,
AND PROFESSOR OF EXPERIMENTAL PHYSICS IN THE UNIVERSITY OF CAMBRIDGE

EDITED BY

WILLIAM GARNETT, M.A.

FORMERLY FELLOW OF ST. JOHN'S COLLEGE, CAMBRIDGE

Oxford

AT THE CLARENDON PRESS

1881

EDITOR'S PREFACE.

MOST of the following pages were written by the late Professor Clerk Maxwell, about seven years ago, and some of them were used by him as the text of a portion of his lectures on Electricity at the Cavendish Laboratory. Very little appears to have been added to the MS. during the last three or four years of Professor Maxwell's life, with the exception of a few fragmentary portions in the latter part of the work. This was partly due to the very great amount of time and thought which he expended upon editing the Cavendish papers, nearly all of which were copied by his own hand, while the experimental investigations which he undertook in order to corroborate Cavendish's results, and the enquiries he made for the purpose of clearing up every obscure allusion in Cavendish's MS., involved an amount of labour which left him very little leisure for other work.

When the MS. came into the hands of the present Editor, the first eight chapters appeared to have been finished and were carefully indexed and the Articles numbered. Chapters IX and X were also provided with tables of contents, but the Articles were not numbered, and several references, Tables, etc., were omitted as well as a few sentences in the text. At the end of the table of contents of Chapter X three points to be treated were mentioned, viz.:—the Passage of Electricity at the surfaces of insulators; Conditions of spark, etc.; Electrification by pressure, friction, rupture, etc.: no Articles corresponding to these headings could be found in the text. Some portions of Chapters IX and X formed separate bundles of MS., and there was no indication of the place which they were intended to fill. This was the case with Arts. 174–181 and 187–192. Arts. 194–196 and 200 also formed a separate MS. with no table of contents and no indication of their intended position.

It was for some time under consideration by the friends of Professor Maxwell, whether the MS. should be published in its fragmentary form or whether it should be completed by another hand, so as to carry out as far as possible the author's original design; but before any decision had been arrived at it was suggested that the book might be made to serve the purposes of students by a selection of Articles from Professor Maxwell's *Electricity and Magnetism*, so as to make it in a sense complete for the portion of the subject covered by the first volume of the last-mentioned work. In accordance with this suggestion, a number of Articles have been selected from the larger book and reprinted. These are indicated by a * after the number of the Article. Arts. 93–98 and 141 are identical with Arts. 118–123 and 58 of the larger treatise, but these have been reprinted in accordance with directions contained in Professor Maxwell's MS.

In the arrangement of the Articles selected from the *Electricity and Magnetism* care has been taken to interfere as little as possible with the continuity of the MS. of the present work, and in some cases logical order has been sacrificed to this object, so that some subjects which are treated briefly in the earlier portions are reintroduced in the latter part of the book. In Chapter XII some articles are introduced from the larger treatise which may appear somewhat inconsistent with the plan of this book; this has been for the sake of the practical value of the results arrived at. The latter part of the note on pages 149 and 150 may be taken as Professor Maxwell's own comment on the method proposed in Art. 186 written a few years subsequently to that Article.

All references, for the accuracy of which Professor Maxwell is not responsible, and all Tables, notes, or interpolations inserted by the Editor, are enclosed in square brackets. This system has not been carried out in the table of contents, but the portion of this contained in Professor Maxwell's MS. is stated above.

Of the Author's Preface the portion here given is all that has been found.

W. G.

CAMBRIDGE,
August, 1881.

FRAGMENT OF AUTHOR'S PREFACE.

THE aim of the following treatise is different from that of my larger treatise on electricity and magnetism. In the larger treatise the reader is supposed to be familiar with the higher mathematical methods which are not used in this book, and his studies are so directed as to give him the power of dealing mathematically with the various phenomena of the science. In this smaller book I have endeavoured to present, in as compact a form as I can, those phenomena which appear to throw light on the theory of electricity, and to use them, each in its place, for the development of electrical ideas in the mind of the reader.

In the larger treatise I sometimes made use of methods which I do not think the best in themselves, but without which the student cannot follow the investigations of the founders of the Mathematical Theory of Electricity. I have since become more convinced of the superiority of methods akin to those of Faraday, and have therefore adopted them from the first.

In the first two chapters experiments are described which demonstrate the principal facts relating to electric charge considered as a quantity capable of being measured.

The third chapter, 'on electric work and energy,' consists of deductions from these facts. To those who have some acquaintance with the elementary parts of mathematics, this chapter may be useful as tending to make their knowledge more precise. Those who are not so prepared may omit this chapter in their first reading of the book.

The fourth chapter describes the electric field, or the region in which electric phenomena are exhibited.

CONTENTS.

CHAPTER I.

CHAPTER II.

ON THE CHARGES OF ELECTRIFIED BODIES.

CHAPTER III.

ON ELECTRICAL WORK AND ENERGY.

CHAPTER IV.

THE ELECTRIC FIELD.

EXPLORATION OF THE ELECTRIC FIELD.

CHAPTER V.

FARADAY'S LAW OF LINES OF INDUCTION.

CHAPTER VI.

PARTICULAR CASES OF ELECTRIFICATION.

CHAPTER VII.

ELECTRICAL IMAGES.

CHAPTER VIII.

CAPACITY.

CHAPTER IX.

ELECTRIC CURRENT.

CHAPTER X.

PHENOMENA OF AN ELECTRIC CURRENT WHICH FLOWS THROUGH HETEROGENEOUS MEDIA.

CHAPTER XI.

METHODS OF MAINTAINING AN ELECTRIC CURRENT.

CHAPTER XII.

ON THE MEASUREMENT OF ELECTRIC RESISTANCE.

CHAPTER XIII.

ON THE ELECTRIC RESISTANCE OF SUBSTANCES.

AN ELEMENTARY TREATISE

ON

ELECTRICITY.

CHAPTER I.

EXPERIMENT I.

Electrification by Friction.

1.] TAKE a stick of sealing-wax, rub it on woollen cloth or flannel, and then bring it near to some shreds of paper strewed on the table. The shreds of paper will move, the lighter ones will raise themselves on one end, and some of them will leap up to the sealing-wax. Those which leap up to the sealing-wax sometimes stick to it for awhile, and then fly away from it suddenly. It appears therefore that in the space between the sealing-wax and the table is a region in which small bodies, such as shreds of paper, are acted on by certain forces which cause them to assume particular positions and to move sometimes from the table to the sealing-wax, and sometimes from the sealing-wax to the table.

These phenomena, with others related to them, are called electric phenomena, the bodies between which the forces are manifested are said to be electrified, and the region in which the phenomena take place is called the electric field.

Other substances may be used instead of the sealing-wax. A piece of ebonite, gutta-percha, resin or shellac will do as well, and so will amber, the substance in which these phenomena were first noticed, and from the Greek name of which the word *electric* is derived.

The substance on which these bodies are rubbed may also be varied, and it is found that the fur of a cat's skin excites them better than flannel.

It is found that in this experiment only those parts of the surface of the sealing-wax which were rubbed exhibit these phenomena, and that some parts of the rubbed surface are apparently more active than others. In fact, the distribution of the electrification over the surface depends on the previous history of the sealing-wax, and this in a manner so complicated that it would be very difficult to investigate it. There are other bodies, however, which may be electrified, and over which the electrification is always distributed in a definite manner. We prefer, therefore, in our experiments, to make use of such bodies.

The fact that certain bodies after being rubbed appear to attract other bodies was known to the ancients. In modern times many other phenomena have been observed, which have been found to be related to these phenomena of attraction. They have been classed under the name of *electric* phenomena, amber, ἤλεκτρον, having been the substance in which they were first described.

Other bodies, particularly the loadstone and pieces of iron and steel which have been subjected to certain processes, have also been long known to exhibit phenomena of action at a distance. These phenomena, with others related to them, were found to differ from the electric phenomena, and have been classed under the name of magnetic phenomena, the loadstone, μάγνης, being found in Magnesia*.

These two classes of phenomena have since been found to be related to each other, and the relations between the various phenomena of both classes, so far as they are known, constitute the science of Electromagnetism.

Experiment II.

Electrification of a Conductor.

2.] Take a metal plate of any kind (a tea-tray, turned upside down, is convenient for this purpose) and support it on three dry wine glasses. Now place on the table a plate of ebonite, a sheet of thin gutta-percha, or a well-dried sheet of brown paper. Rub it lightly with fur or flannel, lift it up from the table by its edges and place it on the inverted tea-tray, taking care not to touch the tray with your fingers.

* The name of Magnesia has been given to two districts, one in Lydia the other in Thessaly. Both seem to have been celebrated for their mineral products, and several substances have been known by the name of magnesia besides that which modern chemists know by that name. The loadstone, the touchstone, and meerschaum, seem however to have been the principal substances which were called Magnesian and magnetic, and these are generally understood to be Lydian stones.

It will be found that the tray is now electrified. Shreds of paper or gold-leaf placed below it will fly up to it, and if the knuckle is brought near the edge of the tray a spark will pass between the tray and the knuckle, a peculiar sensation will be felt, and the tray will no longer exhibit electrical phenomena. It is then said to be *discharged*. If a metal rod, held in the hand, be brought near the tray the phenomena will be nearly the same. The spark will be seen and the tray will be discharged, but the sensation will be slightly different.

If, however, instead of a metal rod or wire, a glass rod, or stick of sealing-wax, or a piece of gutta-percha, be held in the hand and brought up to the tray there will be no spark, no sensation, and no discharge. The discharge, therefore, takes place through metals and through the human body, but not through glass, sealing-wax, or gutta-percha. Bodies may therefore be divided into two classes: conductors, or those which transmit the discharge, and non-conductors, through which the discharge does not take place.

In electrical experiments, those conductors, the charge of which we wish to maintain constant, must be supported by non-conducting materials. In the present experiment the tray was supported on wine glasses in order to prevent it from becoming discharged. Pillars of glass, ebonite, or gutta-percha may be used as supports, or the conductor may be suspended by a white silk thread. Solid non-conductors, when employed for this purpose, are called *insulators*. Copper wires are sometimes lapped with silk, and sometimes enclosed in a sheath of gutta-percha, in order to prevent them from being in electric communication with other bodies. They are then said to be insulated.

The metals are good conductors; air, glass, resins, gutta-percha, caoutchouc, ebonite, paraffin, &c., are good insulators; but, as we shall find afterwards, all substances resist the passage of electricity, and all substances allow it to pass though in exceedingly different degrees. For the present we shall consider only two classes of bodies, good conductors, and good insulators.

EXPERIMENT III.

Positive and Negative Electrification.

3.] Take another tray and insulate it as before, then after discharging the first tray remove the electrified sheet from it and place it on the second tray. It will be found that both trays are now electrified. If a small ball of elder pith suspended by a white silk thread* be made to touch the first tray, it will be immediately repelled from it but attracted towards the second. If it is now allowed to touch the second tray it will be repelled from it but attracted towards the first. The electrifications of the two trays are therefore of opposite kinds, since each attracts what the other repels. If a metal wire, attached to an ebonite rod, be made to touch both trays at once, both trays will be completely discharged. If two pith balls be used, then if both have been made to touch the same tray and then hung up near each other they are found to repel each other, but if they have been made to touch different trays they attract each other. Hence bodies when electrified in the same way are repelled from each other, but when they are electrified in opposite ways they are attracted to each other.

If we distinguish one kind of electrification by calling it *positive*, we must call the other kind of electrification *negative*. We have no physical reason for assigning the name of positive to one kind of electrification rather than to the other. All scientific men, however, are in the habit of calling that kind of electrification positive which the surface of polished glass exhibits after having been rubbed with zinc amalgam spread on leather. This is a matter of mere convention, but the convention is a useful one, provided we remember that it is a convention as arbitrary as that adopted in the diagrams of analytical geometry of calling horizontal distances positive or negative according as they are measured towards the right or towards the left of the point of reference.

In our experiment with a sheet of gutta-percha excited by flannel, the electrification of the sheet and of the tray on which it is placed is negative: that of the flannel and of the tray from which the gutta-percha has been removed is positive.

In whatever way electrification is produced it is one or other of these two kinds.

EXPERIMENT IV.

The Electrophorus of Volta.

4.] This instrument is very convenient for electrical experiments and is much more compact than any other electrifying apparatus.

* I find it convenient to fasten the other end of the thread to a rod of ebonite about 3 mm. diameter. The ebonite is a much better insulator than the silk thread especially in damp weather.

It consists of two disks, and an insulating handle which can be screwed to the back of either plate. One of these disks consists of resin or of ebonite in front supported by a metal back. In the centre of the disk is a metal pin*, which is in connection with the metal back, and just reaches to the level of the surface of the ebonite. The surface of the ebonite is electrified by striking it with a piece of flannel or cat's fur. The other disk, which is entirely of metal, is then brought near the ebonite disk by means of the insulating handle. When it comes within a certain distance of the metal pin, a spark passes, and if the disks are now separated the metal disk is found to be charged positively, and the disk of ebonite and metal to be charged negatively.

In using the instrument one of the disks is kept in connection with one conductor while the other is applied alternately to the first disk and to the other conductor. By this process the two conductors will become charged with equal quantities of electricity, that to which the ebonite disk was applied becoming negative, while that to which the plain metal disk was applied becomes positive.

Electromotive Force.

5.] Definition.—*Whatever produces or tends to produce a transfer of Electrification is called Electromotive Force.*

Thus when two electrified conductors are connected by a wire, and when electrification is transferred along the wire from one body to the other, the tendency to this transfer, which existed before the introduction of the wire, and which, when the wire is introduced, produces this transfer, is called the Electromotive Force *from* the one body *to* the other *along* the path marked out by the wire.

To define completely the electromotive force from one point to another, it is necessary in general to specify a particular path from the one point to the other along which the electromotive force is to be reckoned. In many cases, some of which will be described when we come to electrolytic, thermoelectric, and electromagnetic phenomena, the electromotive force from one point to another may be different along different paths. If we restrict our attention,

* [This was introduced by Professor Phillips to obviate the necessity of touching the carrier plate while in contact with the ebonite.]

however, as we must do in this part of our subject, to the theory of the equilibrium of electricity at rest, we shall find that the electromotive force from one point to another is the same for all paths drawn in air from the one point to the other.

Electric Potential.

6.] The electromotive force from any point, along a path drawn in air, to a certain point chosen as a point of reference, is called the Electric Potential at that point.

Since electrical phenomena depend only on differences of potential, it is of no consequence what point of reference we assume for the zero of potential, provided that we do not change it during the same series of measurements.

In mathematical treatises, the point of reference is taken at an infinite distance from the electrified system under consideration. The advantage of this is that the mathematical expression for the potential due to a small electrified body is thus reduced to its simplest form.

In experimental work it is more convenient to assume as a point of reference some object in metallic connection with the earth, such as any part of the system of metal pipes conveying the gas or water of a town.

It is often convenient to assume that the walls, floor and ceiling of the room in which the experiments are carried on has conducting power sufficient to reduce the whole inner surface of the room to the same potential. This potential may then be taken for zero. When an instrument is enclosed in a metallic case the potential of the case may be assumed to be zero.

Potential of a Conductor.

7.] If the potentials at different points of a uniform conductor are different there will be an electric current from the places of high to the places of low potential. The theory of such currents will be explained afterwards (Chap. ix). At present we are dealing with cases of electric equilibrium in which there are no currents. Hence in the cases with which we have now to do the potential at every point of the conductor must be the same. This potential is called the potential of the conductor.

The potential of a conductor is usually defined as the potential of any point in the air infinitely close to the surface of the conductor. Within a nearly closed cavity in the conductor the potential at any point in the air is the same, and by making the experimental determination of the potential within such a cavity we get rid of the difficulty of dealing with points infinitely close to the surface.

8.] It is found that when two different metals are in contact and in electric equilibrium their potentials as thus defined are in general different. Thus, if a copper cylinder and a zinc cylinder are held in contact with one another, and if first the copper and then the zinc cylinder is made to surround the flame of a spirit lamp, the lamp being in connection with an electrometer, the lamp rapidly acquires the potential of the air within the cylinder, and the electrometer shews that the potential of the air at any point within the zinc cylinder is higher than the potential of the air within the copper cylinder. We shall return to this subject again, but at present, to avoid ambiguity, we shall suppose that the conductors with which we have to do consist all of the same metal at the same temperature, and that the dielectric medium is air.

9.] The region of space in which the potential is higher than a certain value is divided from the region in which it is lower than this value by a surface called an equipotential surface, at every point of which the potential has the given value.

We may conceive a series of equipotential surfaces to be described, corresponding to a series of potentials in arithmetical order. The potential of any point will then be indicated by its position in the series of equipotential surfaces.

No two different equipotential surfaces can cut one another, for no point can have two different potentials.

10.] The idea of electric potential may be illustrated by comparing it with pressure in the theory of fluids and with temperature in the theory of heat.

If two vessels containing the same or different fluids are put into communication by means of a pipe, fluid will flow from the vessel in which the pressure is greater into that in which it is less till the pressure is equalized. This however will not be the case if one vessel is higher than the other, for gravity has a tendency to make the fluid pass from the higher to the lower vessel.

Similarly when two electrified bodies are put into electric communication by means of a wire, electrification will be transferred from the body of higher potential to the body of lower potential, unless there is an electromotive force tending to urge electricity from one of these bodies to the other, as in the case of zinc and copper above mentioned.

Again, if two bodies at different temperatures are placed in thermal communication either by actual contact or by radiation, heat will be transferred from the body at the higher temperature to the body at the lower temperature till the temperature of the two bodies becomes equalized.

The analogy between temperature and potential must not be assumed to extend to all parts of the phenomena of heat and electricity. Indeed this analogy breaks down altogether when it is applied to those cases in which heat is generated or destroyed.

We must also remember that temperature corresponds to a real physical state, whereas potential is a mere mathematical quantity, the value of which depends on the point of reference which we may choose. To raise a body to a high temperature may melt or volatilize it. To raise a body, together with the vessel which surrounds it, to a high potential produces no physical effect whatever on the body. Hence the only part of the phenomena of electricity and heat which we may regard as analogous is the condition of the transfer of heat or of electricity, according as the temperature or the potential is higher in one body or in the other.

With respect to the other analogy—that between potential and fluid pressure—we must remember that the only respect in which electricity resembles a fluid is that it is capable of flowing along conductors as a fluid flows in a pipe. And here we may introduce once for all the common phrase *The Electric Fluid* for the purpose of warning our readers against it. It is one of those phrases, which, having been at one time used to denote an observed fact, was immediately taken up by the public to connote a whole system of imaginary knowledge. As long as we do not know whether positive electricity, or negative, or both, should be called a substance or the absence of a substance, and as long as we do not know whether the velocity of an electric current is to be measured by hundreds of thousands of miles in a second or by an hundreth of an inch in an hour, or even whether the current flows from positive to negative or in the reverse direction, we must avoid speaking of the electric fluid.

On Electroscopes.

11.] An electroscope is an instrument by means of which the existence of electrification may be detected. All electroscopes are capable of indicating with more or less accuracy not only the existence of electrification, but its amount. Such indications, however, though sometimes very useful in guiding the experimenter, are not to be regarded as furnishing a numerical measurement of the electrification. Instruments so constructed that their indications afford data for the numerical measurement of electrical quantities are called Electrometers.

An electrometer may of course be used as an electroscope if it is sufficiently sensitive to indicate the electrification in question, and an instrument intended for an electroscope may, if its indications are sufficiently uniform and regular, be used as an electrometer.

The class of electroscopes of simplest construction is that in which the indicating part of the instrument consists of two light bodies suspended side by side, which, when electrified, repel each other, and indicate their electrification by separating from each other.

The suspended bodies may be balls of elder pith, gilt, and hung up by fine linen threads (which are better conductors than silk or cotton), or pieces of straw or strips of metal, and in the latter case the metal may be tinfoil or gold-leaf, thicker or thinner according to the amount of electrification to be measured.

We shall suppose that our electroscope is of the most delicate kind, in which gold leaves are employed (see Fig. 1). The indicating apparatus *l*, *l*, is generally fastened to one end of a rod of metal *L*, which passes through an opening in the top of a glass vessel *G*. It then hangs within the vessel, and is protected from currents of air which might produce a motion of the suspended bodies liable to be mistaken for that due to electrification.

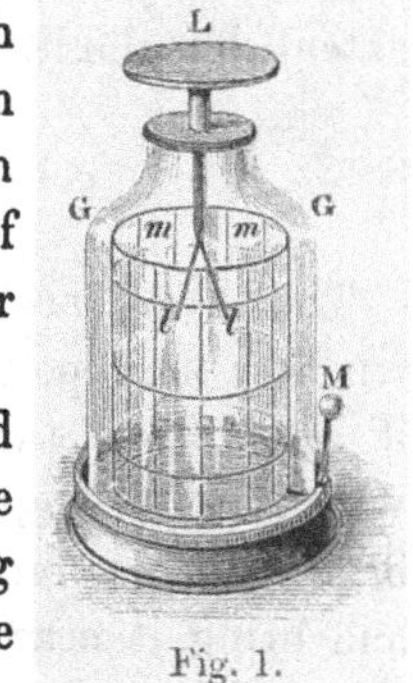

Fig. 1.

To test the electrification of a body the electrified body is brought near the disk *L* at the top of the metal rod, when, if the electrification is strong enough, the suspended bodies diverge from one another.

The glass case, however, is liable, as Faraday pointed out, to become itself electrified, and when glass is electrified it is very difficult to ascertain by experiment the amount and the distribution of its electrification. There is thus introduced into the experiment a new force, the nature and amount of which is unknown, and this interferes with the other forces acting on the gold leaves, so that their divergence can no longer be taken as a true indication of their electrification.

The best method of getting rid of this uncertainty is to place within the glass case a metal vessel which almost surrounds the gold leaves, this vessel being connected with the earth. When the gold leaves are electrified the inside of this vessel, it is true, becomes oppositely electrified, and so increases the divergence of the gold leaves, but the distribution of this electrification is always strictly dependent on that of the gold leaves, so that the divergence of the gold leaves is still a true indication of their actual electrical state. A continuous metal vessel, however, is opaque, so that the gold leaves cannot be seen from the outside. A wire cage, however, may be used, and this is found quite sufficient to shield the gold leaves from the action of the glass, while it does not prevent them from being seen.

The disk, L, and the upper part of the rod which supports the gold leaves, and another piece of metal M, which is connected with the cage m, m, and extends beyond the case of the instrument, are called the *electrodes*, a name invented by Faraday to denote the *ways* by which the electricity gets to the vital parts of the instrument.

The divergence of the gold leaves indicates that the potential of the gold leaves and its electrode is different from that of the surrounding metal cage and its electrode. If the two electrodes are connected by a wire the whole instrument may be electrified to any extent, but the leaves will not diverge.

EXPERIMENT V.

The divergence of the gold leaves does not of itself indicate whether their potential is higher or lower than that of the cage; it only shews that these potentials are different. To ascertain which has the higher potential take a rubbed stick of sealing wax, or any other substance which we know to be negatively electrified, and bring it near the electrode which carries the gold leaves. If the gold leaves are negatively electrified they will diverge more as the sealing wax approaches the rod which carries them; but if they are positively electrified they will tend to collapse. If the electrification of the sealing wax is considerable with respect to that of the gold leaves they will first collapse entirely, but will again open out as the sealing wax is brought nearer, shewing that they are now negatively electrified. If the electrode *M* belonging to the cage is insulated from the earth, and if the sealing wax is brought near it, the indications will be exactly reversed ; the leaves, if electrified positively, will diverge more, and if electrified negatively, will tend to collapse.

If the testing body used in this experiment is positively electrified, as when a glass tube rubbed with amalgam is employed, the indications are all reversed.

By these methods it is easy to determine whether the gold leaves are positively or negatively electrified, or, in other words, whether their potential is higher or lower than that of the cage.

12.] If the electrification of the gold leaves is considerable the electric force which acts on them becomes much greater than their weight, and they stretch themselves out towards the cage as far as they can. In this state an increase of electrification produces no visible effect on the electroscope, for the gold leaves cannot diverge more. If the electrification is still further increased it often happens that the gold leaves are torn off from their support, and the instrument is rendered useless. It is better, therefore, when we have to deal with high degrees of electrification to use a less delicate instrument. A pair of pith balls suspended by linen threads answers very well ; the threads answer sufficiently well as conductors of electricity, and the balls are repelled from each other when electrified.

For very small differences of potential, electroscopes much more sensitive than the ordinary gold-leaf electroscope may be used.

Thomson's Quadrant Electrometer.

13.] In Sir William Thomson's Quadrant Electrometer the indicating part consists of a thin flat strip of aluminium (see Fig. 2) called the needle, attached to a vertical axle of stout platinum wire. It is hung up by two silk fibres x, y, so as to be capable of turning about a vertical axis under the action of the electric force, while it always tends to return to a definite position of equilibrium. The axis carries a concave mirror t by which the image of a flame, and of a vertical wire bisecting the flame, is thrown upon a graduated scale, so as to indicate the motion of the needle round a vertical axis. The lower end of the axle, dips into sulphuric acid contained in the lower part of the glass case of the instrument, and thus puts the needle into electrical connection with the acid. The lower end of the axle also carries a piece of platinum, immersed in the acid which serves to check the vibrations of the needle. The needle hangs within a shallow cylindrical brass box, with circular apertures in the centre of its top and bottom. This box is divided into four quadrants, *a*, *b*, *c*, *d*, which are separately mounted on glass stems, and thus insulated from the case and from one another. The quadrant *b* is removed in the figure to shew the needle. The position of the needle, when in equilibrium, is such, that one end is half in the quadrant *a* and half in *c*, while the other end is half in *b* and half in *d*.

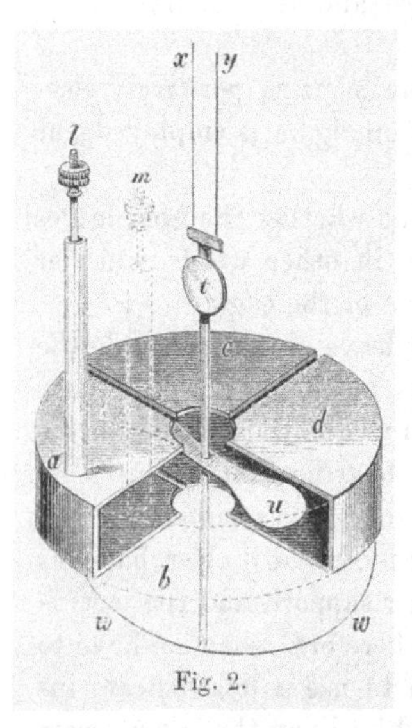

Fig. 2.

The electrode *l* is connected with the quadrant *a* and also with *d* through the wire *w*. The other electrode, *m*, is connected with the quadrants *b* and *c*.

The needle, *u*, is kept always at a high potential, generally positive. To test the difference of potential between any body and the earth, one of the electrodes, say *m*, is connected to the earth, and the other, *l*, to the body to be tested.

The quadrants *b* and *c* are therefore at potential zero, the quadrants *a* and *d* are at the potential to be tested, and the needle *u* is at a high positive potential.

The whole surface of the needle is electrified positively, and the whole inner surface of the quadrants is electrified negatively, but the greatest electrification and the greatest attraction is, other things being equal, where the difference of potentials is greatest. If, therefore, the potential of the quadrants *a* and *d* is positive, the needle will move from *a* and *d* towards *b* and *c* or in the direction of the hands of a watch. If the potential of *a* and *d* is negative, the needle will move towards these quadrants, or in the opposite direction to that of the hands of a watch.

The higher the potential of the needle, the greater will be the force tending to turn the needle, and the more distinct will be the indications of the instrument.

Idiostatic and Heterostatic Instruments.

14.] In the gold-leaf electroscope, the only electrification in the field is the electrification to be tested. In the Quadrant Electrometer the needle is kept always charged. Instruments in which the only electrification is that which we wish to test, are called Idiostatic. Those in which there is electrification independent of that to be tested are called Heterostatic. In an idiostatic instrument, like the gold-leaf electroscope, the indications are the same, whether the potential to be tested is positive or negative, and the amount of the indication is, when very small, nearly as the square of the difference of potential. In a heterostatic instrument, like the quadrant electrometer, the indication is to the one side or to the other, as the potential is positive or negative, and the amount of the indication is proportional to the difference of potentials, and not to the square of that difference. Hence an instrument on the heterostatic principle, not only indicates of itself whether the potential is positive or negative, but when the potential is very small its motion for a small variation of potential is as great as when the potential is large, whereas in the gold leaf electroscope a very small electrification does not cause the gold leaves to separate sensibly.

In Thomson's Quadrant Electrometer there is a contrivance by which the potential of the needle is adjusted to a constant value, and there are other organs for special purposes, which are not represented in the figure which is a mere diagram of the most essential parts of the instrument.

On Insulators.

15.] In electrical experiments it is often necessary to support an electrified body in such a way that the electricity may not escape. For this purpose, nothing is better than to set it on a stand supported by a glass rod, provided the surface of the glass is quite dry. But, except in the very driest weather, the surface of the glass has always a little moisture condensed on it. For this reason electrical apparatus is often placed before a fire, before it is to be used, so that the moisture of the air may not condense on the warmed surface of the glass. But if the glass is made too warm, it loses its insulating power and becomes a good conductor.

Hence it is best to adopt a method by which the surface of the glass may be kept dry without heating it.

The insulating stand in the figure consists of a glass vessel *C*, with a boss rising up in the middle to which is cemented the glass pillar *aa*. To the upper part of this pillar is cemented the neck of the bell glass *B*, which is thus supported so that its rim is within the vessel *C*, but does not touch it. The pillar *a* carries the stand *A* on which the body to be insulated is placed.

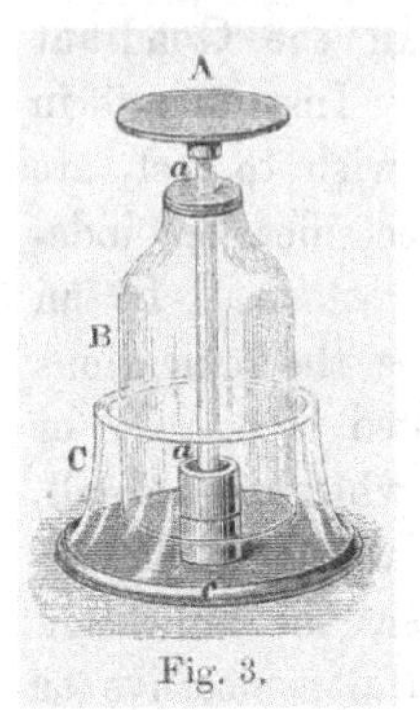

Fig. 3.

In the vessel *C* is placed some strong sulphuric acid *c*, which fills a wide shallow moat round the boss in the middle. The air within the bell glass *B*, in contact with the pillar *a*, is thus dried, and before any damp air can enter this part of the instrument, it must pass down between *C* and *B* and glide over the surface of the sulphuric acid, so that it is thoroughly dried before it reaches the glass pillar. Such an insulating stand is very valuable when delicate experiments have to be performed. For rougher purposes insulating stands may be made with pillars of glass varnished with shellac or of sealing-wax or ebonite.

16.] For carrying about an electrified conductor, it is very convenient to fasten it to the end of an ebonite rod. Ebonite, however, is very easily electrified. The slightest touch with the hand, or friction of any kind, is sufficient to render its surface so electrical, that no conclusion can be drawn as to the electrification of the conductor at the end of the rod.

The rod therefore must never be touched but must be carried by means of a handle of metal, or of wood covered with tinfoil, and before making any experiment the whole surface of the ebonite must be freed from electrification by passing it rapidly through a flame.

The sockets by which the conductors are fastened to the ebonite rods, should not project outwards from the conductors, for the electricity not only accumulates on the projecting parts, but creeps over the surface of the ebonite, and remains there when the electricity of the conductor is discharged. The sockets should therefore be entirely within the outer surface of the conductors as in Fig. 4.

It is convenient to have a brass ball, (Fig. 4), one inch in diameter, a cylindrical metal vessel (Fig. 5) about three inches in diameter and five or six inches deep, a pair of disks of tin plate (Figs. 6, 7), two inches in diameter, and a thin wire about a foot long (Fig. 8) to make connection between electrified bodies. These should all be mounted on ebonite rods (penholders), one eighth of an inch diameter, with handles of metal or of wood covered with tinfoil.

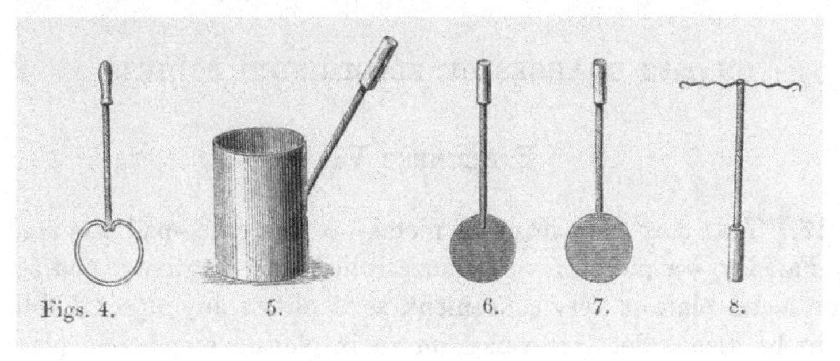

Figs. 4. 5. 6. 7. 8.

CHAPTER II.

ON THE CHARGES OF ELECTRIFIED BODIES.

Experiment VI.

17.] Take any deep vessel of metal,—a pewter ice-pail was used by Faraday,—a piece of wire gauze rolled into a cylinder and set on a metal plate is very convenient, as it allows any object within it to be seen. Set this vessel on an insulating stand, and place an electroscope near it. Connect one electrode of the electroscope permanently with the earth or the walls of the room, and the other with the insulated vessel, either permanently by a wire reaching from the one to the other, or occasionally by means of a wire carried on an ebonite rod and made to touch the vessel and the electrode at the same time. We shall generally suppose the vessel in permanent connection with the electroscope. The simplest way when a gold leaf electroscope is used is to set the vessel on the top of it.

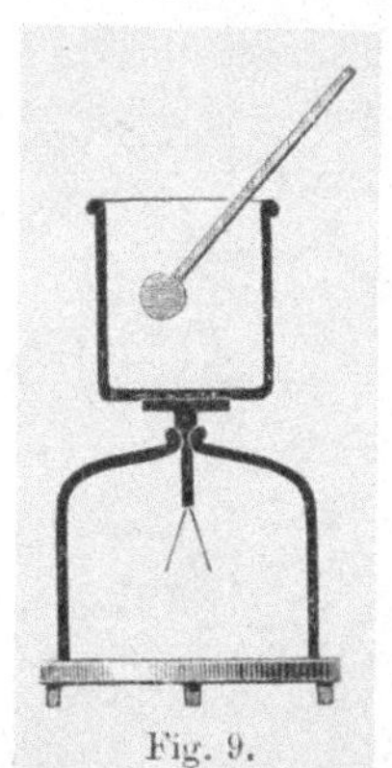
Fig. 9.

Take a metal ball at the end of an ebonite rod, electrify it by means of the electrophorus, and carrying it by the rod as a handle let it down into the metal vessel without touching the sides.

As the electrified ball approaches the vessel the indications of the electroscope continually increase, but when the ball is fairly within the vessel, that is when its depth below the opening of the vessel becomes considerable compared with the diameter of the opening, the indications of the electroscope no longer increase, but remain unchanged in whatever way the ball is moved about within the vessel.

This statement, which is approximately true for any deep vessel, is rigorously true for a closed vessel. This may be shewn by closing the mouth of the vessel with a metal lid worked by means of a silk thread. If the electrified ball be drawn up and let down in the vessel by means of a silk thread passing through a hole in the lid, the external electrification of the vessel as indicated by the electrometer will remain unchanged, while the ball changes its position within the vessel. The electrification of the gold leaves when tested is found to be of the *same* kind as that of the ball.

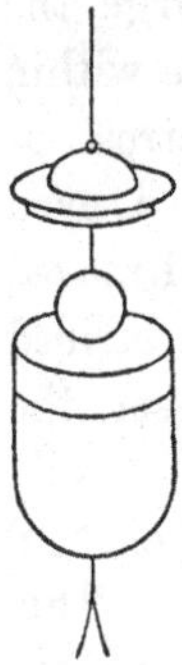

Fig. 10.

Now touch the outside of the vessel with the finger, so as to put it in electric communication with the floor and walls of the room. The external electrification of the vessel will be discharged, and the gold leaves of the electroscope will collapse. If the ball be now moved about within the vessel, the electroscope will shew no signs of electrification; but if the ball be taken out of the vessel without touching the sides, the gold leaves will again diverge as much as they did during the first part of the experiment. Their electrification however will be found to be of the *opposite* kind from that of the ball.

Experiment VII.

To compare the charges or total Electrification of two electrified balls.

18.] Since whatever be the position of the electrified bodies within the vessel its external electrification is the same, it must depend on the total electrification of the bodies within it, and not on the distribution of that electrification. Hence, if two balls, when alternately let down into the vessel, produce the same divergence of the gold leaves, their charges must be equal. This may be further tested by discharging the outside of the vessel when the first ball is in it, and then removing it and letting the second ball down into the vessel. If the charges are equal, the electroscope will still indicate no electrification.

If we wish to ascertain whether the charges of two bodies, oppositely electrified, are numerically equal, we may do so by discharging the vessel and then letting down both bodies into it. If the charges are equal and opposite, the electroscope will not be affected.

EXPERIMENT VIII.

When an electrified body is hung up within a closed metallic vessel, the total electrification of the inner surface of the vessel is equal and opposite to that of the body.

19.] After hanging the body within the vessel, discharge the external electrification of the vessel, and hang up the whole within a larger vessel connected with the electroscope. The electroscope will indicate no electrification, and will remain unaffected even if the electrified body be taken out of the smaller vessel and moved about within the larger vessel. If, however, either the electrified body or the smaller vessel be removed from the large vessel, the electroscope will indicate positive or negative electrification.

When an electrified body is placed within a vessel free of charge, the external electrification is equal to that of the body. This follows from the fact already proved that the internal electrification is equal and opposite to that of the body, and from the circumstance that the total charge of the vessel is zero.

But it may also be proved experimentally by placing, first the electrified body itself, and then the electrified body surrounded by an uncharged vessel, within the larger vessel and observing that the indications of the electroscope are the same in both cases.

EXPERIMENT IX.

When an electrified body is placed within a closed vessel and then put into electrical connection with the vessel, the body is completely discharged.

20.] In performing any of the former experiments bring the electrified body into contact with the inside of the vessel, and then take it out and test its charge by placing it within another vessel connected with the electroscope. It will be found quite free of charge. This is the case however highly the body may have been originally electrified, and however highly the vessel itself, the inside of which it is made to touch, may be electrified.

If the vessel, during the experiment, is kept connected with the electroscope, no alteration of the external electrification can be detected at the moment at which the electrified body is made to touch the inside of the vessel. This affords another proof that the electrification of the interior surface is equal and opposite to that of the electrified body within it. It also shews that when there is no electrified body within the surface every part of that surface is free from charge.

EXPERIMENT X.

To charge a vessel with any number of times the charge of a given electrified body.

21.] Place a smaller vessel within the given vessel so as to be insulated from it. Place the electrified body within the inner vessel, taking care not to discharge it. The exterior charges of the inner and outer vessels will now be equal to that of the body, and their interior charges will be numerically equal but of the opposite kind. Now make electric connection between the two vessels. The exterior charge of the inner vessel and the interior charge of the outer vessel will neutralise each other, and the outer vessel will now have a charge equal to that of the body, and the inner vessel an equal and opposite charge.

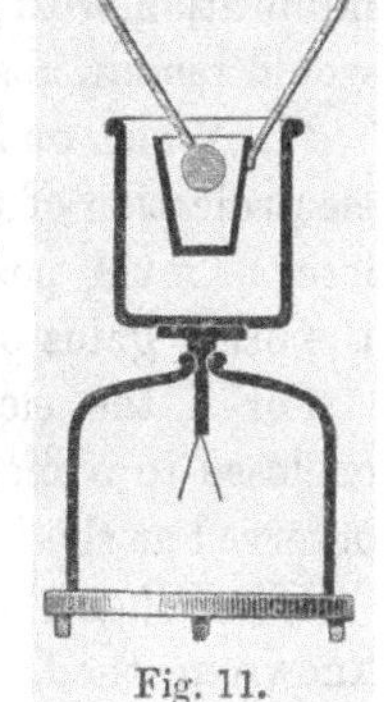
Fig. 11.

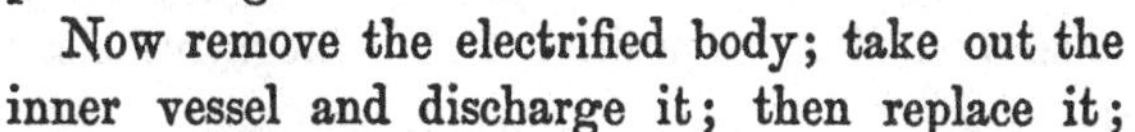

Now remove the electrified body; take out the inner vessel and discharge it; then replace it; place the electrified body within it; and make contact between the vessels. The outer vessel has now received a double charge, and by repeating this process any number of charges, each equal to that of the electrified body, may be communicated to the outer vessel.

To charge the outer vessel with electrification opposite to that of the electrified body is still easier. We have only to place the electrified body within the smaller vessel, to put this vessel for a moment in connection with the walls of the room so as to discharge the exterior electrification, then to remove the electrified body and carry the vessel into the inside of the larger vessel and bring it into contact with it so as to give the larger vessel its negative charge, and then to remove the smaller vessel, and to repeat this process the required number of times.

We have thus a method of comparing the electric charges of different bodies without discharging them, of producing charges equal to that of a given electrified body, and either of the same or of opposite signs, and of adding any number of such charges together.

22.] In this way we may illustrate and test the truth of the following laws of electrical phenomena.

I. The total electrification or charge of a body or system of bodies remains always the same, except in so far as it receives electrification from, or gives electrification to other bodies.

In all electrical experiments the electrification of bodies is found to change, but it is always found that this change arises from defective insulation, and that as the means of insulation are improved, the loss of electrification becomes less. We may therefore assert that the electrification of a body cut off from electrical communication with all other bodies by a perfectly insulating medium would remain absolutely constant.

II. When one body electrifies another by conduction the total electrification of the two bodies remains the same, that is, the one loses as much positive or gains as much negative electrification as the other gains of positive or loses of negative electrification.

For if the electric connection is made when both bodies are enclosed in a metal vessel, no change of the total electrification is observed at the instant of contact.

III. When electrification is produced by friction or by any other known method, equal quantities of positive and of negative electricity are produced.

For if the process of electrification is conducted within the closed vessel, however intense the electrification of the parts of the system may be, the electrification of the whole, as indicated by the electroscope connected with the vessel, remains zero.

IV. If an electrified body or system of bodies be placed within a closed conducting surface (which may consist of the floor, walls, and ceiling of the room in which the experiment is made), the interior electrification of this surface is equal and opposite to the electrification of the body or system of bodies.

V. If no electrified body is placed within the hollow conducting surface, the electrification of this surface is zero. This is true, not only of the electrification of the surface as a whole, but of every part of this surface.

For if a conductor be placed within the surface and in contact with it, the surface of this conductor becomes electrically continuous with the interior surface of the enclosing vessel, and it is found that if the conductor is removed and tested, its electrification is always zero, shewing that the electrification of every part of an interior surface within which there is no electrified body is zero.

By means of Thomson's Quadrant Electrometer it is easy to measure the electrification of a body when it is a million times less than when charged to an amount convenient for experiment. Hence the experimental evidence for the above statements shews that they cannot be erroneous to the extent of one-millionth of the principal electrifications concerned.

CHAPTER III.

ON ELECTRICAL WORK AND ENERGY.

23.] Work in general is the act of producing a change of configuration in a material system in opposition to a force which resists this change.

Energy is the capacity of doing work.

When the nature of a material system is such that if after the system has undergone any series of changes it is brought back in any manner to its original state, the whole work done by external agents on the system is equal to the whole work done by the system in overcoming external forces, the system is called a Conservative system.

The progress of physical science has led to the investigation of different forms of energy, and to the establishment of the doctrine, that all material systems may be regarded as conservative systems provided that all the different forms of energy are taken into account. This doctrine, of course, considered as a deduction from experiment, can assert no more than that no instance of a non-conservative system has hitherto been discovered, but as a scientific or science-producing doctrine it is always acquiring additional credibility from the constantly increasing number of deductions which have been drawn from it, which are found in all cases to be verified. In fact, this doctrine is the one generalised statement which is found to be consistent with fact, not in one physical science only, but in all. When once apprehended it furnishes to the physical enquirer a principle on which he may hang every known law relating to physical actions, and by which he may be put in the way to discover the relations of such actions in new branches of science. For such reasons the doctrine is commonly called the Principle of the Conservation of Energy.

General Statement of the Conservation of Energy.

24.] The total energy of any system of bodies is a quantity which can neither be increased nor diminished by any mutual action of those bodies, though it may be transformed into any of the forms of which energy is susceptible.

If, by the action of some external agent, the configuration of the system is changed, then, if the forces of the system are such as to *resist* this change of configuration, the external agent is said to do work on the system. In this case the energy of the system is *increased*. If, on the contrary, the forces of the system tend to *produce* the change of configuration, so that the external agent has only to *allow* it to take place, the system is said to do work on the external agent, and in this case the energy of the system is diminished. Thus when a fish has swallowed the angler's hook and swims off, the angler following him for fear his line should break, the fish is doing work against the angler, but when the fish becomes tired and the angler draws him to shore, the angler is doing work against the fish.

Work is always measured by the product of the change of configuration into the force which resists that change. Thus, when a man lifts a heavy body, the change of configuration is measured by the increase of distance between the body and the earth, and the force which resists it is the weight of the body. The product of these measures the work done by the man. If the man, instead of lifting the heavy body vertically upwards, rolls it up an inclined plane to the same height above the ground, the work done against gravity is precisely the same; for though the heavy body is moved a greater distance, it is only the vertical component of that distance which coincides in direction with the force of gravity acting on the body.

25.] If a body having a positive charge of electricity is carried by a man from a place of low to a place of high potential, the motion is opposed by the electric force, and the man does work on the electric system, thereby increasing its energy. The amount of work is measured by the product of the number of units of electricity into the increase of potential in moving from the one place to the other.

We thus obtain the dynamical definition of electric potential.

The electric potential at a given point of the field is measured by the amount of work which must be done by an external agent in carrying one unit of positive electricity from a place where the potential is zero to the given point.

This definition is consistent with the imperfect definition given at Art. 6, for the work done in carrying a unit of electricity from one place to another will be positive or negative according as the displacement is from lower to higher or from higher to lower potential. In the latter case the motion, if not prevented, will take place, without any interference from without, in obedience to the electric forces of the system. Hence the flow of electricity along conductors is always from places of high to places of low potential.

26.] We have already defined the electromotive force from one place to another along a given path as the work done by the electric forces of the system on a unit of electricity carried along that path. It is therefore measured by the excess of the potential at the beginning over that at the end of the path.

The electromotive force *at a point* is the force with which the electrified system would act on a small body electrified with a unit of positive electricity, and placed at that point.

If the electrified body is moved in such a way as to remain on the same equipotential surface, no work is done by the electric forces or against them. Hence the direction of the electric force acting on the small body is such that any displacement of the body along any line drawn on the equipotential surface is at right angles to the force. The direction of the electromotive force, therefore, is at right angles to the equipotential surface.

The magnitude of this force, multiplied by the distance between two neighbouring equipotential surfaces, gives the work done in passing from the one equipotential surface to the other, that is to say, the difference of their potentials.

Hence the magnitude of the electric force may be found by dividing the difference of the potentials of two neighbouring equipotential surfaces by the distance between them, the distance being, of course, very small, and measured perpendicularly to either surface. The direction of the force is that of the normal to the equipotential surface through the given point, and is reckoned in the direction in which the potential *diminishes*.

INDICATOR DIAGRAM OF ELECTRIC WORK.

27.] The indicator diagram, employed by Watt for measuring the work done by a steam engine,* may be made use of in investigating the work done during the charging of a conductor with electricity.

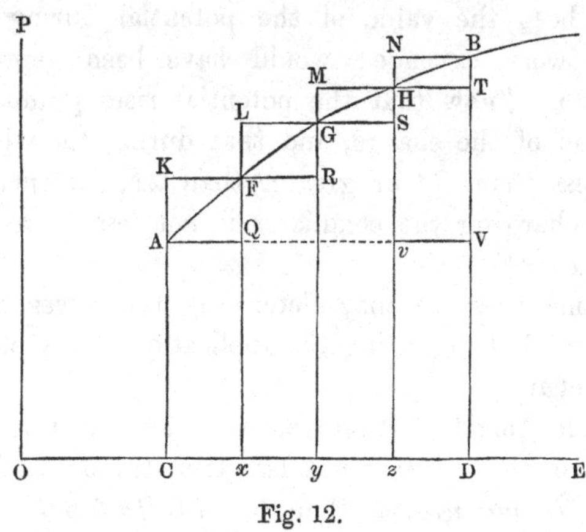

Fig. 12.

Let the charge of the conductor at any instant be represented by a horizontal line OC, drawn from the point O, which is called the *origin* of the diagram, and let the potential of the conductor at the same instant be represented by a vertical line CA, drawn from the extremity of the first line, then the position of the extremity of the second line will indicate the electric state of the conductor, both with respect to its charge, and also with respect to its potential.

If during any electrical operation this point moves along the line $AFGHB$, we know not only that the charge has been increased from the value OC to the value OD, and that the potential has been increased from CA to DB, but that the charge and the potential at any instant, corresponding, say, to the point F of the curve, are represented respectively by Ox and xF.

28.] *Theorem.* The work expended by an external agent in bringing the increment of charge from the walls of the room to the conductor is represented by the area enclosed by the base line CD, the two vertical lines CA and DB, and the curve $AFGHB$.

For let CD, the increment of the charge, be divided into any number of equal parts at the points x, y, z.

* Maxwell's 'Theory of Heat,' 4th ed., p. 102.

The value of the potential just before the application of the charge Cx is represented by AC. Hence if the potential were to remain constant during the application of the charge Cx, the work expended in charging the conductor would be represented by the product of this potential and the charge, or by the area $ACxQ$.

As soon as the charge Cx has been applied the potential is xF. If this had been the value of the potential during the whole process, the work expended would have been represented by $KCxF$. But we know that the potential rises gradually during the application of the charge, and that during the whole process it is never less than CA or greater than xF. Hence the work expended in charging the conductor is not less than $ACxQ$, nor greater than $KCxF$.

In the same way we may determine the lower and higher limits of the work done during the application of any other portion of the entire charge.

We conclude, therefore, that the work expended in increasing the charge from OC to OD is not less than the area of the figure $CAQFRGSHTD$, nor greater than $CKFLGMHNBD$. The difference between these two values is the sum of the parallelograms KQ, LR, MS, NT, the breadths of which are equal, and their united height is BV. Their united area is therefore equal to that of the parallelogram $NvVB$.

By increasing without limit the number of equal parts into which the charge is divided, the breadth of the parallelograms will be diminished without limit. In the limit, therefore, the difference of the two values of the work vanishes, and either value becomes ultimately equal to the area $CAFGHBD$, bounded by the curve, the extreme ordinates, and the base line.

This method of proof is applicable to any case in which the potential is always increasing or always diminishing as the charge increases. When this is not the case, the process of charging may be divided into a number of parts, in each of which the potential is either always increasing or always diminishing, and the proof applied separately to each of these parts.

Superposition of Electric Effects.

29.] It appears from Experiment VII that several electrified bodies placed in a hollow vessel produce each its own effect on the electrification of the vessel, in whatever positions they are placed.

From this it follows that one electric phenomenon at least, that called electrification by induction, is such that the effect of the whole electrification is the sum of the effects due to the different parts of the electrification. The different electrical phenomena, however, are so intimately connected with each other that we are led to infer that all other electrical phenomena may be regarded as composed of parts, each part being due to a corresponding part of the electrification.

Thus if a body A, electrified in a definite manner, would produce a given potential, P, at a given point of the field, and if a body, B, also electrified in a definite manner, would produce a potential, Q, at the same point of the field, then when both bodies, still electrified as before, are introduced simultaneously into their former places in the field, the potential at the given point will be $P+Q$. This statement may be verified by direct experiment, but its most satisfactory verification is founded on a comparison of its consequences with actual phenomena.

As a particular case, let the electrification of every part of the system be multiplied n fold. The potential at every point of the system will also be multiplied by n.

30.] Let us now suppose that the electrical system consists of a number of conductors (which we shall call A_1, A_2, &c.) insulated from each other, and capable of being charged with electricity. Let the charges of these conductors be denoted by E_1, E_2, &c., and their potentials by P_1, P_2, &c.

If at first the conductors are all without charge, and at potential zero, and if at a certain instant each conductor begins to be charged with electricity, so that the charge increases uniformly with the time, and if this process is continued till the charges simultaneously become E_1 for the first conductor, E_2 for the second, and so on, then since the increment of the charge of any conductor is the same for every equal interval of time during the process, the increment of the potential of each conductor will also be the same for every equal increment of time, so that the line which represents, on the indicator diagram, the succession of states of a given conductor with respect to charge and potential will be described with

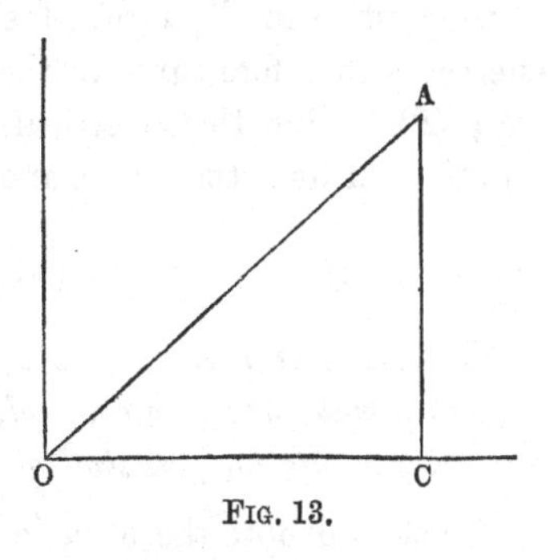

FIG. 13.

a velocity, the horizontal and vertical components of which remain constant during the process. This line on the diagram is therefore a straight line, drawn from the origin, which represents the initial state of the system when devoid of charge and at potential zero, to the point A_1 which indicates the final state of the conductor when its charge is E, and its potential P_1, and will represent the process of charging the conductor A_1. The work expended in charging this conductor is represented by the area OCA, or half the product of the final charge E and the final potential P.

Energy of a System of Electrified Bodies.

31.] When the relative positions of the conductors are fixed, the work done in charging them is entirely transformed into electrical energy. If they are charged in the manner just described, the work expended in charging any one of them is $\frac{1}{2}EP$, where E represents its final charge and P its final potential. Hence the work expended in charging the whole system may be written

$$\tfrac{1}{2}E_1P_1 + \tfrac{1}{2}E_2P_2 + \&c.,$$

there being as many products as there are conductors in the system.

It is convenient to write the sum of such a series of terms in the form

$$\tfrac{1}{2}\Sigma(EP),$$

where the symbol Σ (sigma) denotes that all the products of the form EP are to be summed together, there being one such product for each of the conductors of which the system consists.

Since an electrified system is subject to the law of Conservation of Energy, the work expended in charging it is entirely stored up in the system in the form of electrical energy. The value of this energy is therefore equal to that of the work which produced it, or $\frac{1}{2}\Sigma(EP)$. But the electrical energy of the system depends altogether on its actual state, and not on its previous history. Hence

Theorem I.

The electrical energy of a system of conductors, in whatever way they may have been charged, is half the sum of the products of the charge into the potential of each conductor.

We shall denote the electric energy of the system by the symbol Q, where

$$Q = \tfrac{1}{2}\Sigma(EP). \qquad (1)$$

Work done in altering the charges of the system.

32.] We shall next suppose that the conductors of the system, instead of being originally without charge and at potential zero, are originally charged with quantities E_1, E_2, &c. of electricity, and are at potentials P_1, P_2, &c. respectively.

When in this state let the charges of the conductors be changed, each at a uniform rate, the rate being, in general, different for each conductor, and let this process go on uniformly, till the charges have become E_1', E_2', &c., and the potentials P_1', P_2', &c. respectively.

By the principle of the superposition of electrical effects the increment of the potential will vary as the increment of the charge, and the potential of each conductor will increase or diminish at a uniform rate from P to P', while its charge varies at a uniform rate from E to E'. Hence the line AA', which represents the varying state of the conductor during the process, is the straight line which joins A, the point which indicates its original state, with A', which represents its final state. The work spent in producing this increment of charge in the conductor is represented by the area $ACC'A'$, or $\frac{1}{2}CC'\,(CA+C'A')$, or $(E'-E)\,\frac{1}{2}(P+P')$, or, in words, it is the product of the increase of charge and the half sum of the potentials at the beginning and end of the operation, and this will be true for every conductor of the system.

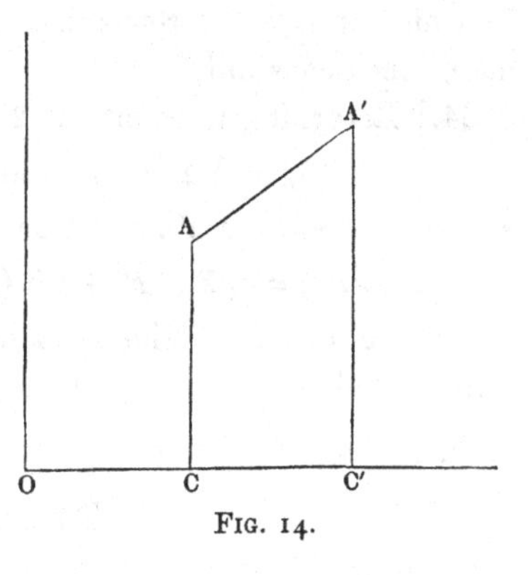

FIG. 14.

As, during this process, the electric energy of the system changes from Q, its original, to Q', its final value, we may write

$$Q' = Q + \tfrac{1}{2}\Sigma\{(E'-E)(P'+P)\}, \quad \text{.............. (2)}$$

hence,

THEOREM II.

The increment of the energy of the system is half the sum of the products of the increment of charge of each conductor into the sum of its potentials at the beginning and the end of the process.

33.] If all the charges but one are maintained constant (by the insulation of the conductors) the equation (2) is reduced to

$$Q'-Q = (E'-E)\,\tfrac{1}{2}(P'+P),$$

or
$$\frac{Q'-Q}{E'-E} = \tfrac{1}{2}(P'+P). \qquad (3)$$

If the increment of the charge is taken always smaller and smaller till it ultimately vanishes, P' becomes equal to P and the equation may be interpreted thus :—

The *rate* of increase of the electrical energy due to the increase of the charge of one of the conductors at a rate unity is numerically equal to the potential of that conductor.

In the notation of the differential calculus this result is expressed by the equation
$$\frac{dQ_e}{dE} = P, \qquad (4)$$
in which it is to be remembered that all the charges but one are maintained constant.

34.] Returning to equation (2), we have already shewn that
$$Q = \tfrac{1}{2}\Sigma(EP) \quad \text{and} \quad Q' = \tfrac{1}{2}\Sigma(E'P'); \qquad (5)$$
we may therefore write equation (2)
$$\tfrac{1}{2}\Sigma(E'P') = \tfrac{1}{2}\Sigma(EP)+\tfrac{1}{2}\Sigma(E'P'-EP+E'P-EP'). \qquad (6)$$

Cutting out from the equation the terms which destroy each other, we obtain
$$\Sigma(EP') = \Sigma(E'P), \qquad (7)$$
or in words,

THEOREM III.

In a fixed system of conductors the sum of the products of the original charge and the final potential of each conductor is equal to the sum of the products of the final charge and the original potential.

This theorem corresponds, in the elementary treatment of electrostatics, to Green's Theorem in the analytical theory. By properly choosing the original and the final state of the system we may deduce a number of results which we shall find useful in our after-work.

35.] In the first place we may write, as before,
$$\tfrac{1}{2}\Sigma\{(E'-E)(P'+P)\} = \tfrac{1}{2}\Sigma(E'P'-EP+E'P-EP'); \qquad (8)$$
adding and subtracting the equal quantities of equation (7),
$$O = \Sigma(EP'-E'P), \qquad (9)$$
and the right-hand side becomes
$$\tfrac{1}{2}\Sigma(E'P'-EP-E'P+EP'), \qquad (10)$$

or $\frac{1}{2}\Sigma\{(E'-E)(P'+P)\} = Q'-Q = \frac{1}{2}\Sigma\{(E'+E)(P'-P)\}$, (11)

or in words,

THEOREM IV.

The increment of the energy of a fixed system of conductors is equal to half the sum of the products of the increment of the potential of each conductor into the sum of the original and final charges of that conductor.

36.] If all the conductors but one are maintained at constant potentials (which may be done by connecting them with voltaic batteries of constant electromotive force), equation (11) is reduced to

$$Q'-Q = \tfrac{1}{2}(E'+E)(P'-P), \quad (12)$$

or

$$\frac{Q'-Q}{P'-P} = \tfrac{1}{2}(E'+E). \quad (13)$$

If the increment of the potential is taken successively smaller and smaller, till it ultimately vanishes, E' becomes at last equal to E and the equation may be interpreted thus :—

The *rate* of increase of the electrical energy due to the increase of potential of one of the conductors at a rate unity is numerically equal to the charge of that conductor.

In the notation of the differential calculus this result is expressed by the equation

$$\frac{dQ_p}{dP} = E, \quad (14)$$

in which it is to be remembered that all the potentials but one are maintained constant.

37.] We have next to point out some of the results which may be deduced from Theorem III.

If any conductor, as A_t, is insulated and without charge both in the initial and the final state, then $E_t = 0$ and $E_t' = 0$, and therefore

$$E_t P_t = 0 \text{ and } E_t' P_t = 0, \quad (15)$$

so that the terms depending on A_t disappear from both members of equation (7).

Again, if another conductor, say A_u, be connected with the earth both in the initial and in the final state, $P_u = 0$ and $P_u' = 0$, so that

$$E_u P_u' = 0 \text{ and } E_u' P_u = 0\,;$$

so that, in this case also, the terms depending on A_u disappear from both sides of equation (7).

If, therefore, all the conductors with the exception of two, say

A_r and A_s, are either insulated and without charge, or else connected with the earth, equation (7) is reduced to the form

$$E_r P_r' + E_s P_s' = E_r' P_r + E_s' P_s. \quad \text{..............} \quad (16)$$

Let us first suppose that in the initial state all the conductors except A_r are without charge, and that in the final state all the conductors except A_s are without charge. The equation then becomes

$$E_r P_r' = E_s' P_s, \quad \text{......................} \quad (17)$$

or

$$\frac{P_s}{E_r} = \frac{P_r'}{E_s'},$$

or in words,

THEOREM V.

In a system of fixed insulated conductors, the potential (P_s) produced in A_s by a charge E communicated to A_r is equal to the potential (P_r') produced in A_r by an equal charge E communicated to A_s.

This is the first instance we have met with of the *reciprocal* relation of two bodies. There are many such reciprocal relations. They occur in every branch of science, and they often enable us to deduce the solution of new electrical problems from those of simpler problems with which we are already familiar. Thus, if we know the potential which an electrified sphere produces at a point in its neighbourhood, we can deduce the effect which a small electrified body, placed at that point, would have in raising the potential of the sphere.

38.] Let us next suppose that the original potential of A_s is P_s and that all the other conductors are kept at potential zero by being connected with the walls of the room, and let the final potential of A_r be P_r', that of all the others being zero, then in equation (7) all the terms involving zero potentials will vanish, and we shall have in this case also

$$E_r P_r' = E_s' P_s. \quad \text{........................} \quad (18)$$

If, therefore, $\quad P_r' = P_s, \quad E_r = E_s', \quad \text{....................} \quad (19)$

or in words,

THEOREM VI.

In a system of fixed conductors, connected, all but one, with the walls of the room, the charge (E_r) induced on A_r when A_s is raised to the potential P_s is equal to the charge (E_s') induced on A_s when A_r is raised to an equal potential (P_r').

39.] As a third case, let us suppose all the conductors insulated and without charge, and that a charge is communicated to A_r

which raises its potential to P_r and that of A_s to P_s. Next, let A_s be connected with the earth, and let a charge E_r' on A_r induce the charge E_s' on A_s.

In equation (16) we have $E_r = 0$ and $P_s' = 0$, so that the left-hand member vanishes and the equation becomes

$$0 = E_r' P_r + E_s' P_s, \qquad (20)$$

or

$$\frac{P_s}{P_r} = -\frac{E_r'}{E_s'}.$$

Hence, if $\qquad P_s = n P_r, \qquad E_r' = -n E_s', \qquad (21)$

or in words,

THEOREM VII.

If in a system of fixed conductors insulated and originally without charge a charge be communicated to A_r which raises its potential to P_r, unity, and that of A_s to n, then if in the same system of conductors a charge unity be communicated to A_s and A_r be connected with the earth the charge induced on A_r will be $-n$.

If, instead of supposing the other conductors A_t &c. to be all insulated and without charge, we had supposed some or all of them to be connected with the earth, the theorem would still be true, only the value of n would be different according to the arrangement we adopt.

This is one of Green's theorems. As an example of its application, let us suppose that we have ascertained the distribution of electric charge induced on the various parts of the surface of a conductor by a small electrified body in a given position with unit charge. Then by means of this theorem we can solve the following problem. The potential at every point of a surface coinciding in position with that of the conductor being given, determine the potential at a point corresponding to the position of the small electrified body.

Hence, if the potential is known at all points of any closed surface, it may be determined for any point within that surface if there be no electrified body within it, and for any point outside if there be no electrified body outside.

Mechanical work done by the electric forces during the displacement of a system of insulated electrified conductors.

40.] Let A_1, A_2 &c. be the conductors, E_1, E_2 &c. their charges, which, as the conductors are insulated, remain constant. Let P_1, P_2 &c. be their potentials before and P_1', P_2' &c. their potentials after the displacement. The electrical energy of the system before the displacement is

$$Q = \tfrac{1}{2}\Sigma(EP). \quad \text{......................}(22)$$

During the displacement the electric forces which act in the same direction as the displacement perform an amount of work equal to W, and the energy remaining in the system is

$$Q' = \tfrac{1}{2}\Sigma(EP'). \quad \text{......................}(23)$$

The original energy, Q, is thus transformed into the work W and the final energy Q', so that the equation of energy is

$$Q = W + Q', \quad \text{..........................}(24)$$

or

$$W = \tfrac{1}{2}\Sigma[E(P-P')]. \quad \text{..................}(25)$$

This expression gives the work done during any displacement, small or large, of an insulated system. To find the force, we must make the displacement so small that the configuration of the system is not sensibly altered thereby. The ultimate value of the quotient found by dividing the work by the displacement is the value of the force resolved in the direction of the displacement.

Mechanical work done by the electric force during the displacement of a system of conductors each of which is kept at a constant potential.

41.] Let us begin by supposing each conductor of the system insulated, and that a *small* displacement is given to the system, by which the potential is changed from P to P_1. The work done during this displacement is, as we have shewn,

$$W = \tfrac{1}{2}\Sigma[E(P-P_1)]. \quad \text{....................}(26)$$

Next, let the conductors remain fixed while the charges of the conductors are altered from E to E_1, so as to bring back the value of the potential from P_1 to P. Then we know by equation (7) that

$$\Sigma(EP - E_1P_1) = 0. \quad \text{.......................}(27)$$

Hence, substituting for $\Sigma(EP)$ in (26),

$$W = \tfrac{1}{2}\Sigma[(E_1-E)P_1]. \quad \text{...................}(28)$$

Performing these two operations alternately for any number of times, and distinguishing each pair of operations by a suffix, we find the whole work

$$W = W_1 + W_2 + \&c. \quad \text{..}(29)$$

$$= \tfrac{1}{2}\Sigma[(E_1-E)P_1] + \tfrac{1}{2}\Sigma[(E_2-E_1)P_2] + \&c. \quad \text{.....}(30)$$

By making each of the partial displacements small enough, the values of P_1, P_2 &c. may be made to approach without limit to P, the constant value of the potential, and the expression becomes

$$W = \tfrac{1}{2}\Sigma[(E_1 - E)P] + \tfrac{1}{2}\Sigma[(E_2 - E_1)P] + \text{\&c.} + \tfrac{1}{2}\Sigma[(E' - E_{n-1})P], \quad (31)$$

where E' is the value of E after the last operation. The final result is therefore

$$W = \tfrac{1}{2}\Sigma[(E' - E)P], \quad (32)$$

which is an expression giving the work done during a displacement of any magnitude of a system of conductors, the potential of each of which is maintained constant during the displacement.

We may write this result

$$W = \tfrac{1}{2}\Sigma(E'P) - \tfrac{1}{2}\Sigma(EP), \quad (33)$$

or

$$W = Q' - Q; \quad (34)$$

or the work done by the electric forces is equal to the *increase* of the electric energy of the system during the displacement when the *potential* of each conductor is maintained constant. When the *charge* of each conductor was maintained constant, the work done was equal to the *decrease* of the energy of the system.

Hence, when the potential of each conductor is maintained constant during a displacement in which a quantity of work, W, is done, the voltaic batteries which are employed to keep the potentials constant must do an amount of work equal to $2W$. Of this energy supplied to the system, half is spent in increasing the energy of the system, and the other half appears as mechanical work.

CHAPTER IV.

THE ELECTRIC FIELD.

42.] We have seen that, when an electrified body is enclosed in a conducting vessel, the total electrification of the interior surface of the surrounding vessel is invariably equal in numerical value but opposite in kind to that of the body. This is true, however large this vessel may be. It may be a room of any size having its floor, walls and ceiling of conducting matter. Its boundaries may be removed further, and may consist of the surface of the earth, of the branches of trees, of clouds, perhaps of the extreme limits of the atmosphere or of the universe. In every case, wherever we find an electrified insulated body, we are sure to find at the boundaries of the insulating medium, wherever they may be, an equal amount of electrification of the opposite kind.

This correspondence of properties between the two limits of the insulating medium leads us to examine the state of this medium itself, in order to discover the reason why the electrification at its inner and outer boundaries should be thus related. In thus directing our attention to the state of the insulating medium, rather than confining it to the inner conductor and the outer surface, we are following the path which led Faraday to many of his electrical discoveries.

43.] To render our conceptions more definite, we shall begin by supposing a conducting body electrified positively and insulated within a hollow conducting vessel. The space between the body and the vessel is filled with air or some other insulating medium. We call it an *insulating* medium when we regard it simply as retaining the charge on the surface of the electrified body. When we consider it as taking an important part in the manifestation of electric phenomena we shall use Faraday's expression, and call it a *dielectric* medium. Finally, when we contemplate the region occupied by the medium as being a part of space in which electric phenomena may be observed, we shall call this region the Electric Field. By using this last expression we are not obliged to figure to ourselves the mode in which the dielectric medium takes part in the phenomena. If we afterwards wish to form a theory of the action of the medium, we may find the term dielectric useful.

Exploration of the Electric Field.

Experiment XI.

(*a*) *Exploration by means of a small electrified body.*

44.] Electrify a small round body, a gilt pith ball, for example, and carry it by means of a white silk thread into any part of the field. If the ball is suspended in such a way that the tension of the string exactly balances the weight of the ball, then when the ball is placed in the electric field it will move under the action of a new force developed by the action of the electrified ball on the electric condition of the field. This new force tends to move the ball in a certain direction, which is called the direction of the electromotive force.

If the charge of the ball is varied, the force is sensibly proportional to the charge, provided this charge is not sufficient to produce a sensible disturbance of the state of electrification of the system. If the charge is positive, the force which acts on the ball is, on the whole, *from* the positively electrified body, and *towards* the negatively electrified walls of the room. If the charge is negative, the force acts in the opposite direction.

Since, therefore, the force which acts on the ball depends partly on the charge of the ball and partly on its position and on the electrification of the system, it is convenient to regard this force as the product of two factors, one being the charge of the ball, and the other *the electromotive force at that point of the field which is occupied by the centre of the ball.*

This electromotive force at the point is definite in magnitude and direction. A positively charged body placed there tends to move in the positive direction of the line representing the force. A negatively charged body tends to move in the opposite direction.

EXPERIMENT XII.

(β) *Exploration by means of two disks.*

45.] But the electromotive force not only tends to move electrified bodies, it also tends to transfer electrification from one part of a body to another.

Take two small equal thin metal disks, fastened to handles of shellac or ebonite; discharge them and place them face to face in the electric field, with their planes perpendicular to the direction of the electromotive force. Bring them into contact, then separate them and remove them, and test first one and then the other by introducing them into the hollow vessel of Experiment VII. It will be found that each is charged, and that if the electromotive force acts in the direction AB, the disk on the side of A will be charged negatively, and that on the side of B positively, the numerical values of these charges being equal. This shews that there has been an actual transference of electricity from the one disk to the other, the direction of this transference being that of the electromotive force. This experiment with two disks affords a much more convenient method of measuring the electromotive force at a point than the experiment with the charged pith ball. The measurement of small forces is always a difficult operation, and becomes almost impossible when the weight of the body acted on forms a disturbing force and has to be got rid of by the adjustment of counterpoises. The measurement of the charges of the disks, on the other hand, is much more simple.

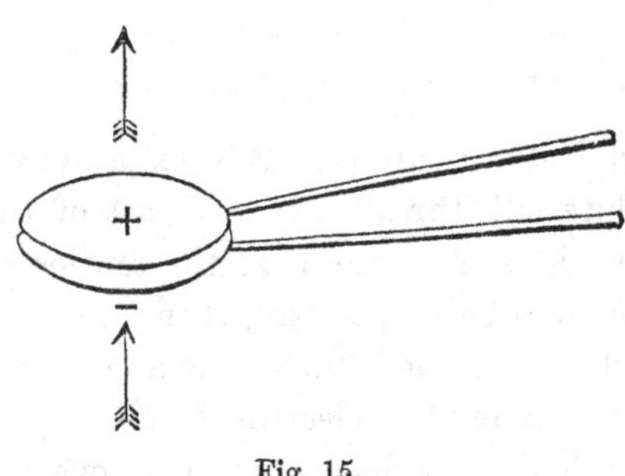

Fig. 15.

The two disks, when in contact, may be regarded as forming a single disk, and the fact that when separated they are found to have received equal and opposite charges, shews that while the disks were in contact there was a distribution of electrification between them, the electrification of each disk being opposite to that of the body next to it, whether the insulated body, which is charged positively, or the inner surface of the surrounding vessel, which is charged negatively.

Electric Tension.

46.] The two disks, after being brought into contact, tend to separate from each other, and to approach the oppositely electrified surfaces to which they are opposed. The force with which they tend to separate is proportional to the area of the disks, and it increases as the electromotive force increases, not, however, in the simple ratio of that force, but in the ratio of the square of the electromotive force.

The electrification of each disk is proportional to the electromotive force, and the mechanical force on the disk is proportional to its electrification and the electromotive force conjointly, that is, to the *square* of the electromotive force.

This force may be accounted for if we suppose that at every point of the dielectric at which electromotive force exists there is a tension, like the tension of a stretched rope, acting in the direction of the electromotive force, this tension being proportional to the square of the electromotive force at the point. This tension acts only on the outer side of each disk, and not on the side which is turned towards the other disk, for in the space between the disks there is no electromotive force, and consequently no tension.

The expression Electric Tension has been used by some writers in different senses. In this treatise we shall always use it in the sense explained above,—the tension of so many pounds' or grains' weight on the square foot exerted by the air or other dielectric medium in the direction of the electromotive force.

Experiment XIII.

Coulomb's Proof Plane.

47.] If one of these disks be placed with one of its flat surfaces in contact with the surface of an electrified conductor and then removed, it will be found to be charged. If the disk is very thin, and if the electrified surface is so nearly flat that the whole surface of the disk lies very close to it, the charge of the disk will be nearly equal to that of the portion of the electrified surface which it covered. If the disk is thick, or does not lie very close to the electrified surface, its charge, when removed, will be somewhat greater.

This method of ascertaining the density of electrification of a surface was introduced by Coulomb, and the disk when used for this purpose is called Coulomb's Proof Plane.

The charge of the disk is by Experiment XII proportional to the electromotive force at the electrified surface. Hence the electromotive force close to a conducting surface is proportional to the density of the electrification at that part of the surface.

Since the surface of the conductor is an equipotential surface, the electromotive force is perpendicular to it. The fact that the electromotive force at a point close to the surface of a conductor is perpendicular to the surface and proportional to the density of the electrification at that point was first established experimentally by Coulomb, and it is generally referred to as Coulomb's Law.

To prove that when the proof plane coincides with the surface of the conductor the charge of the proof plane when removed from the electrified conductor is equal to the charge on the part of the surface which it covers, we may make the following experiment.

A sphere whose radius is 5 units is placed on an insulating stand. A segment of a thin spherical shell is fastened to an insulating handle. The radius of the spherical surface of the shell is 5, the diameter of the circular edge of the segment is 8, and the height of the segment is 2. When applied to the sphere it covers one-fifth part of its surface. A second sphere, whose radius is also 5, is placed on an insulating handle.

The first sphere is electrified, the segment is then placed in contact with it and removed. The second sphere is then made to touch the first sphere, removed and discharged, and then made to touch the first sphere again. The segment is then placed within a conducting vessel, which is discharged to earth, and then insulated and the segment removed. One of the spheres is then made to touch the outside of the vessel, and is found to be perfectly discharged.

Let e be the electrification of the first sphere, and let the charge removed by the segment be ne, then the charge remaining on the sphere is $(1-n)e$. The charge of the first sphere is then divided with the second sphere, and becomes $\frac{1}{2}(1-n)e$. The second sphere is then discharged, and the charge is again divided, so that the charge on either sphere is $\frac{1}{4}(1-n)e$. The charge on the insulated vessel is equal and opposite to that on the segment, and it is therefore $-ne$, and this is perfectly neutralized by the charge on one of the spheres; hence $\frac{1}{4}(1-n)e+(-ne)=0$,

from which we find $n=\frac{1}{5}$,

or the electricity removed by the segment covering one-fifth of the surface of the sphere is one-fifth of the whole charge of the sphere.

Experiment XIV.

Direction of Electromotive Force at a Point.

48.] A convenient way of determining the direction of the electromotive force is to suspend a small elongated conductor with its middle point at the given point of the field. The two ends of the short conductor will become oppositely electrified, and will then be drawn in opposite directions by the electromotive force, so that the axis of the conductor will place itself in the direction of the force at that point. A short piece of fine cotton or linen thread, through the middle of which a fine silk fibre is passed, answers very well. The silk fibre, held by both ends, serves to carry the piece of thread into any desired position, and the thread then takes up the direction of the electric force at that place.

Experiment XV.

Potential at any Point of the Field.

49.] Suspend two small uncharged metal balls in the field by silk threads, and then connect them by means of a fine metal wire fastened to the end of an ebonite rod. Remove the wire and the spheres separately, and then examine the charges of the spheres.

It will be found that the two spheres, if they have become electrified, have received equal and opposite charges. If the potentials at the points of the field occupied by the centres of the spheres are different, positive electrification will be transferred from the place of high to the place of low potential, and the sphere at the place of high potential will thus become charged negatively, and that at the place of low potential will become charged positively. These charges may be shewn to be proportional to the difference of potentials at the two places.

We have thus a method of determining points of the field at which the potential is the same. Place one of the spheres at a fixed point, and move the other about till, on connecting the spheres with a wire as before, no charge is found on either sphere. The potentials of the field at the points occupied by the centres of the spheres must now be the same. In this way a number of points may be found, the potential at each of which is equal to that at a given point.

All these points lie on a certain surface, which is called an equipotential surface. On one side of this surface the potential is higher, on the other it is lower, than at the surface itself.

We have seen that electricity has no tendency to flow from one part of such a surface to another. An electrified body, if constrained so as to be capable of moving only from one point of the surface to another, would be in equilibrium, and the force acting on such a body is therefore everywhere perpendicular to the equipotential surface.

Experiment XVI.

50.] We may use one sphere only, and after placing it with its centre at any given point of the field we may touch it for a moment with a wire connected to the earth. We may then remove the sphere and determine its charge. The charge of the sphere is proportional to the potential at the given point, a positive charge, however, corresponding to a negative potential.

Equipotential Surfaces.

51.] In this way the potential at any number of points in the field may be ascertained, and equipotential surfaces may be supposed drawn corresponding to values of the potential represented by the numbers 1, 2, 3, &c.

These surfaces will form a series, each, in general, lying within the preceding surface and having the succeeding surface within it. No two distinct surfaces can intersect each other, though a particular equipotential surface may consist of two or more sheets, intersecting each other at certain lines or points.

The surface of any conductor in electric equilibrium is an equipotential surface. For if the potential at one point of the conductor is different from that at another point, electricity will flow from the place of higher potential to the place of lower potential till the potentials are rendered equal.

Experiment XVII.

52.] To make an experimental determination of the equipotential surfaces belonging to an electrified system we may use a small exploring sphere permanently connected by a fine wire with one electrode of the electroscope, the other electrode being connected with the earth. Place the centre of the sphere at a given point, and connect the electrodes together for an instant. The indication of the electroscope will thus be reduced to zero. If the sphere is now moved in such a manner that the indication of the electrometer remains zero during the motion, the path of the centre of the exploring sphere will lie on an equipotential surface. For if it moves to a place of higher potential, electricity will flow from the sphere to the electroscope, and if it moves to a place of lower potential, electricity will flow from the electroscope to the sphere.

If the bodies belonging to the electrified system are not perfectly insulated, their potentials and the potentials of the points of the field will tend to approach zero. The path in which the centre of the exploring sphere moves is such that its potential at any point has a given value at the time when the centre of the sphere passes it. The different points of the path are not therefore on a surface which has the same potential at any one instant, for the potential is diminishing everywhere, and the path must therefore pass from surfaces of lower to surfaces of higher potential so as to make up for this loss.

Experiment XVIII.

53.] The following method, founded on Theorem V, Art. 37, is therefore in many cases more convenient, as it is much easier to secure good insulation for the exploring sphere on an insulating handle than for a large electrified conductor of irregular form. Let it be required to determine the equipotential surfaces due to the electrification of the conductor C placed on an insulating stand. Connect the conductor with one electrode of the electroscope, the other being connected with the earth. Electrify the exploring sphere, and,

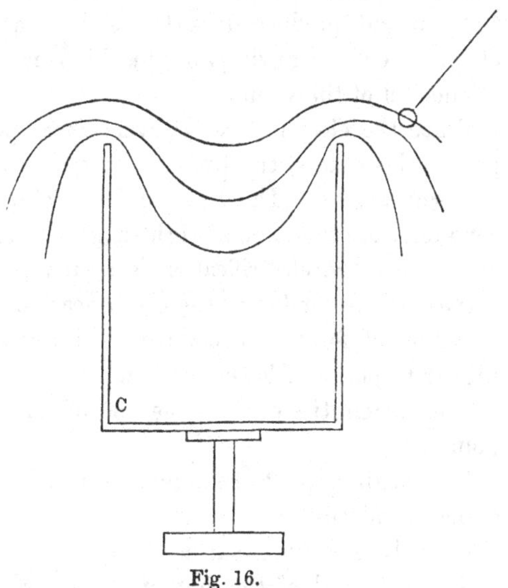

Fig. 16.

carrying it by the insulating handle, bring its centre to a given point. Connect the electrodes for an instant, and then move the sphere in such a path that the indication of the electroscope remains zero. This path will lie on an equipotential surface.

For by Theorem V, the part of the potential of the conductor C due to the presence of the charged exploring sphere with its centre at a given point is equal to the potential at the given point due to a charge on the conductor C equal to that of the exploring sphere.

By this method the potential of the conductor remains zero, or very nearly zero, during the whole time of the experiment, so that there is very little tendency to change of the charge of this body. The exploring sphere, on the other hand, is at a high potential, but as it is not connected by a wire with any other body, its insulation may be made very good.

Lines of Electric Force:

54.] If the direction of the electric force at various points of the field be determined, and if a line be drawn so that its direction at every point of its course coincides with the direction of the electric force at that point, such a line is called a Line of Force. By drawing a number of such lines, the direction of the force at different parts of the field may be indicated.

The lines of force and equipotential surfaces may be drawn, not in the electric field itself, where the mechanical operation of drawing them might produce disturbance, but in a model or plan of the electric field. Drawings of this kind are given in Plates I to VI at the end of the volume.

Since the electric force is everywhere perpendicular to the equipotential surfaces, the lines of force cut these surfaces everywhere at right angles. The lines of force which meet the surface of a conductor are therefore at right angles to it. When they issue from the surface the electrification is positive, and when they enter the surface of the conductor the electrification is negative.

A line of force in every part of its course passes from places of higher to places of lower potential.

The extremities of the same line of force are called *corresponding* points.

The beginning of the line is a point on a positively electrified surface, and the end of the line is a corresponding point on a negatively electrified surface.

The potential of the first of these surfaces must be higher than that of the second.

CHAPTER V.

FARADAY'S LAW OF LINES OF INDUCTION.

55.] FARADAY in his electrical researches employs the lines of force to indicate, not only the direction of the electric force at each point of the field, but also the quantity of electrification on any given portion of the electrified surface.

If we consider a portion of an electrified surface as cut off from the rest by the bounding line which surrounds it, and if from every point of this bounding line we draw a line of force, producing it till it meets the surface of some other body in a point which is said to *correspond* to the point of the body from which the line was drawn, these lines will form a tubular surface, and will cut off a certain portion from the surface of the other body corresponding to the portion of the surface of the first body, and the total electrifications of the two corresponding portions are equal in numerical magnitude but opposite in kind.

56.] A particular instance of Faraday's law is that which we have already proved by experiment, namely, that the electrification of the inner surface of a closed conducting vessel is equal and opposite to that of an electrified body placed within it. Here we have a relation between the whole electrification of the inner surface and that of the opposed surface of the interior body. Faraday's law asserts that, by drawing lines of force from the one surface to the other, points corresponding to each other in the two surfaces may be found; that corresponding lines are such that any point of one has its corresponding point in the other; and that the electrifications of the two portions of the opposed surfaces bounded by such corresponding lines are equal and opposite.

57.] We have called these lines 'lines of force' because we began by defining them as lines whose *direction* at every point coincides with that of the electric force. Every line of force begins at a positively electrified surface and ends at a negatively electrified surface, and the points of these surfaces at which it begins and ends are called *corresponding* points.

A system of lines of force forming a tubular surface closed at the one end by a portion of the positively electrified surface and at the other by the corresponding portion of the negative surface, is called by Faraday a *Tube of Induction*, because electric induction, according to Faraday, is that condition of the dielectric by which the electrifications of the opposed surfaces are placed in that physical relation to one another, which we express by saying that their electrifications are equal and opposite.

Properties of a Tube of Induction.

58.] (1) The electrification of the portion of the positively electrified surface from which the tube of induction proceeds is equal in numerical value but opposite in sign to the negative electrification of the portion of the surface at which the tube of induction terminates.

By dividing the positive surface into portions, the electrification of each of which is unity, and drawing tubes corresponding to each portion, we obtain a system of *unit* tubes, which will be very convenient in describing electric phenomena. For in this case the electrification of any surface is measured by the *number* of tubes which abut on it. If they proceed *from* the surface, this number is to be taken as representing the *positive* electrification; if the tubes terminate at the surface, the electrification is negative.

It is in this sense that Faraday so often speaks of th *number* of lines of force which fall on a given area.

If we suppose an imaginary surface drawn in the electric field, then the quantity of electrostatic induction through this surface is measured by the number of tubes of induction which pass through it, and is reckoned positive or negative accordingly as the tubes pass through it in the positive or negative direction.

Note. By an imaginary surface is meant a surface which has no physical existence, but which may be imagined to exist in space without interfering with the physical properties of the substance which occupies that space. Thus we may imagine a vertical plane dividing a man's head longitudinally into two equal parts, and by means of this imaginary surface we may render our ideas of the form of his head more precise, though any attempt to convert this imaginary surface into a physical one would be criminal. Imaginary quantities, such as are mentioned in treatises on analytical geometry, have no place in physical science.

59.] In every part of the course of a line of electrostatic induction it is passing from places of higher to places of lower potential, and in a direction at right angles to the equipotential surfaces which it cuts.

We have seen that the electric field is divided by the equipotential surfaces into a series of shells, like the coats of an onion, the thickness of each shell at any point being inversely as the electric force at that point.

We have now divided the electric field into a system of unit tubes of induction, the section of each tube at any point varying inversely as the intensity of the electric induction at that point.

Each of these tubes is cut by the equipotential surfaces into a number of segments which we may call unit cells.

60.] If we take the simplest case—that of a single positively electrified body placed within a closed conducting vessel, all the tubes of induction begin at the positively electrified body and end at the negatively electrified surface of the inner vessel. The number of these tubes, since they are unit-tubes, is equal to the number of electrical units in the charge of the electrified body. Each of them cuts all the equipotential surfaces which enclose the electrified body and are enclosed by the vessel. Each tube, therefore, is divided into a number of cells representing the difference of the potential of the electrified body from that of the vessel. If e is the charge of the body and p its potential, E and P being the charge and potential of the vessel, the whole number of cells is

$$e(p-P),$$

or, since $E=-e$, we may write this expression

$$ep+EP.$$

Now this is double of the expression which we formerly obtained for the electrical energy of the system (see Art. 31). Hence in this simple case the number of cells is double the number of units of energy in the system.

If there are several electrified bodies, A, B, C, &c., the tubes of induction proceeding from one of them, A, may abut either on the inner surface of the surrounding vessel or on one of the other electrified bodies.

Let E_1, E_2, E_3 be the charges of A, B, C and P_1, P_2, P_3 their potentials, the charge and potential of the vessel being E_0 and P_0.

Let E_{AB}, E_{AC}, E_{AO} denote the number of tubes of induction which pass from A to the conductors B and C and the vessel O, respectively. Then the whole number of cells will be

$$E_{AB}(P_1-P_2)+E_{AC}(P_1-P_3)+E_{AO}(P_1-P_0),$$
$$+E_{BC}(P_2-P_3)+E_{BO}(P_2-P_0),$$
$$+E_{CO}(P_3-P_0).$$

By arranging the terms according to the potentials involved in them, and remembering that since E_{AB} denotes the number of tubes which pass from A to B, E_{BA} must denote the number which pass from B to A and therefore

$$E_{BA} = -E_{AB},$$

the expression may be written

$$P_1(E_{AB}+E_{AC}+E_{AO}),$$
$$+P_2(E_{BC}+E_{BO}+E_{BA}),$$
$$+P_3(E_{CO}+E_{CA}+E_{CB}),$$
$$+P_0(E_{OA}+E_{OB}+E_{OC}).$$

Now $E_{AB}+E_{AC}+E_{AO}$ is the whole number of tubes issuing from A and this therefore is equal to E_1, the charge of A, and the coefficients of the other potentials are also the charges of the bodies to which they refer, so that the final expression is

$$P_0E_0+P_1E_1+P_2E_2+P_3E_3,$$

and this is double the energy of the system.

Hence, whether there is one electrified body or several, the number of cells is twice the number of units of electrical energy in the system.

61.] This remarkable correspondence between the number of cells into which the tubes of induction are cut by the equipotential surfaces, and the electrical energy of the system, leads us to enquire whether the electrical energy may not have its true seat in the dielectric medium which is thus cut up into cells, each cell being a portion of the medium in which half a unit of energy is stored up. We have only to suppose that the electromotive force, when it acts on a dielectric, puts it into a certain state of constraint, from which it is always endeavouring to relieve itself.

To make our conception of what takes place more precise, let us consider a single cell belonging to a tube of induction proceeding from a positively electrified body, the cell being bounded by two consecutive equipotential surfaces surrounding the body.

We know that there is an electromotive force acting outwards from the electrified body. This force, if it acted on a conducting medium, would produce a current of electricity which would last as long as the force continued to act. The medium however is a non-conducting or dielectric medium, and the effect of the electromotive force is to produce what we may call electric displacement, that is to say, the electricity is forced outwards in the direction of the electromotive force, but its condition when so displaced is such that, as soon as the electromotive force is removed, the electricity resumes the position which it had before displacement.

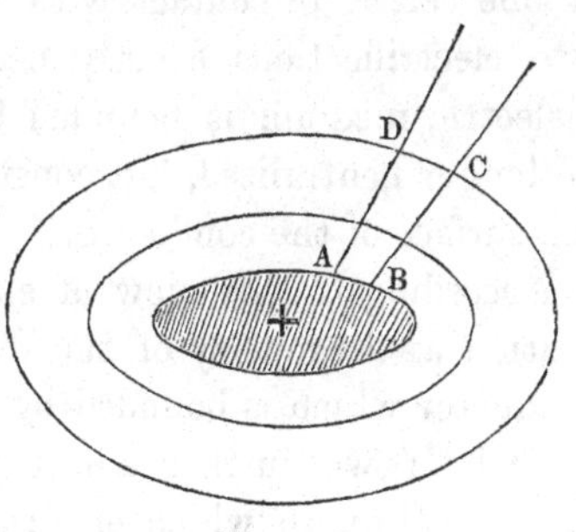

Fig. 17.

The amount of electric displacement is measured by the quantity of electricity which crosses an imaginary fixed surface drawn parallel to the equipotential surfaces.

We know absolutely nothing with respect to the distance through which any particular portion of electricity is displaced from its original position. The only thing we know is the quantity which crosses a given surface. The greater the amount of electricity which we suppose to exist, say, in a cubic inch, the smaller the distance through which we must suppose it displaced in order that a given quantity of electricity may be displaced across a square inch of area fixed in the medium. It is probable that the actual motion of displacement is exceedingly small, in which case we must suppose the quantity of electricity in a cubic inch of the medium to be exceedingly great. If this is really the case the actual velocity of electricity in a telegraph wire may be very small, less, say, than the hundredth of an inch in an hour, though the signals which it transmits may be propagated with great velocity.

62.] The displacement across any section of a unit tube of induction is one unit of electricity and the direction of the displacement is that of the electromotive force, namely, from places of higher to places of lower potential.

Besides the electric displacement within the cell we have to consider the state of the two ends of the cell which are formed by the equipotential surfaces. We must suppose that in every cell the end formed by the surface of higher potential is coated with one unit of positive electricity, the opposite end, that formed by the surface of lower potential, being coated with one unit of negative electricity. In the interior of the medium where the positive end of one cell is in contact with the negative end of the next, these two electrifications exactly neutralise each other, but where the dielectric medium is bounded by a conductor, the electrification is no longer neutralised, but constitutes the observed electrification at the surface of the conductor.

According to this view of electrification, we must regard electrification as a property of the dielectric medium rather than of the conductor which is bounded by it.

63.] If we further admit that in every part of a dielectric medium through which electric induction is taking place there is a tension, like that of a rope, in the direction of the lines of force, and a pressure in all directions at right angles to the lines of force, we may account for all the mechanical actions which take place between electrified bodies.

The tension, referred to unit of surface, is proportional to the square of the electromotive force at the point. The pressure has the same numerical value, but is, of course, opposite in sign.

In my larger treatise on electricity a proof is given of the fact that a system of stress such as is here described is consistent with the equilibrium of a fluid dielectric medium, and that this state of stress in the medium is mechanically equivalent to the attraction or repulsion which electrified bodies manifest.

I have not, however, attempted, by any hypothesis as to the internal constitution of the dielectric medium, to explain in what way the electric displacement causes or is associated with this state of stress.

We have thus, by means of the tubes of induction and the equipotential surfaces, constructed a geometrical model of the field of lectric force. Diagrams of particular cases are given in the figures at the end of this book.

The direction and magnitude of the electric force at any point may be indicated either by means of the equipotential surfaces or by means of the tubes of induction. Hence, when it is expressed in both ways, we may by the study of the relation between the equipotential surfaces and the tubes of induction deduce important theorems in the theory of electricity.

On the use of Physical Analogies.

64.] In many cases the relations of the phenomena in two different physical questions have a certain similarity which enables us, when we have solved one of these questions, to make use of our solution in answering the other. The similarity which constitutes the analogy is not between the phenomena themselves, but between the relations of these phenomena.

To begin with a case of extreme simplicity;—a person slow at arithmetic having to find the price of 52 yards of cotton at 7 pence a yard, if he happened to remember that there are 52 weeks and a day in a year of 365 days, might at once give the answer, 364 pence, without performing the calculation. Here there is no resemblance whatever between the quantities themselves—the weeks and the yards of cotton,—the sole resemblance is between the arithmetical relations of these quantities to others in the same question.

The analogy between electrostatic phenomena and those of the uniform conduction of heat in solid bodies was first pointed out by Sir W. Thomson in a paper 'On the Uniform Motion of Heat in Homogeneous Solid Bodies, and its connection with the Mathematical Theory of Electricity,' published in the *Cambridge Mathematical Journal*, Feb. 1842; reprinted in the *Phil. Mag.* 1854, and in the reprint of Thomson's papers on *Electrostatics and Magnetism.* The analogy is of the following nature:—

Electrostatics.	*Heat.*
The electric field.	An unequally heated body.
A dielectric medium.	A body which conducts heat.
The electric potential at different points of the field.	The temperature at different points in the body.
The electromotive force which tends to move positively electrified bodies from places of higher to places of lower potential.	The flow of heat by conduction from places of higher to places of lower temperature.
A conducting body.	A perfect conductor of heat.
The positively electrified surface of a conductor.	A surface through which heat flows into the body.
The negatively electrified surface of a conductor.	A surface through which heat escapes from the body.
A positively electrified body.	A source of heat.
A negatively electrified body.	A sink of heat, that is, a place at which heat disappears from the body.
An equipotential surface.	An isothermal surface.
A line or tube of induction.	A line or tube of flow of heat.

By a judicious use of this analogy and other analogies of the same nature the progress of physical science has been greatly assisted. In order to avoid the dangers of crude hypotheses we must study the true nature of analogies of this kind. We must not conclude from the partial similarity of some of the relations of the phenomena of heat and electricity that there is any real physical similarity between the causes of these phenomena. The similarity is a similarity between relations, not a similarity between the things related.

This similarity is so complete as far as it goes that any result we may have obtained either about electricity or about the conduction of heat may be at once translated out of the language of the one science into that of the other without fear of error; and in pursuing our investigations in either subject we are at liberty to make use of the ideas belonging to the other, if by so doing we are enabled to see more clearly the connection between one step and another of the reasoning.

We must bear in mind that at the time when Sir W. Thomson pointed out the analogy between electrostatic and thermal phenomena men of science were as firmly convinced that electric attraction was a direct action between distant bodies as that the conduction of heat was the continuous flow of a material fluid through a solid body. The dissimilarity, therefore, between the things themselves appeared far greater to the men of that time than to the readers of this book, who, unless they have been previously instructed, have not yet learned either that heat is a fluid or that electricity acts at a distance.

65.] But we must now consider the limits of the analogy—the points beyond which we must not push it.

In the first place, it is only a particular class of cases of the conduction of heat that have analogous cases in electrostatics. In general, when heat is flowing through a body it causes the temperature of some parts of the body to rise and that of others to fall, and the flow of heat, which depends on the relation of these temperatures, is therefore variable. If the supply of heat is maintained uniform, the temperatures of the different parts of the body tend to adjust themselves to a state in which they remain constant. The quantity of heat which enters any given portion of the body is then exactly equal to that which leaves it during the same time. Under these circumstances the flow of heat is said to be steady.

Now the analogy with electric phenomena applies to the steady flow of heat only. The more general case, that of variable flow of heat, has nothing in electrostatics analogous to it. Even the restricted case of steady flow of heat differs in a most important element from the electrostatic analogue. The steady flow of heat must be kept up by the continual supply of heat at a constant rate and the continual withdrawal of heat at an equal rate. This involves a continual expenditure of energy to maintain the flow of heat in a constant state, so that though the state of the body remains constant and independent of time, the element of time enters into the calculation of the amount of heat required.

The element of time does not enter into the corresponding case in electrostatics. So far as we know, a set of electrified bodies placed in a perfectly insulating medium might remain electrified for ever without a supply of anything from external sources. There is nothing in this case to which we can apply the term 'flow,' which we apply to the case of the transmission of heat with the same propriety that we apply it to the case of a current of electricity, of water, or of time itself.

66.] Another limitation to the analogy is that the temperature of a body cannot be altered without altering its physical state. The density, conductivity, electric properties, &c. all vary when the temperature rises.

The electrical potential, however, which is the analogue of temperature is a mere scientific concept. We have no reason to regard it as denoting a physical state. If a number of bodies are placed within a hollow metallic vessel which completely surrounds them, we may charge the outer surface of the vessel and discharge it as we please without producing any physical effect whatever on the bodies within. But we know that the electric potential of the enclosed bodies rises and falls with that of the vessel. This may be proved by passing a conductor connected to the earth through a hole in the vessel. The relation of the enclosed bodies to this conductor will be altered by charging and discharging the vessel. But if the conductor be removed, the simultaneous rise and fall of the potentials of the bodies in the vessel is not attended with any physical effect whatever.

67.] Faraday* proved this by constructing a hollow cube, twelve feet in the side, covered with good conducting materials, insulated from the ground and highly electrified by a powerful machine. 'I went into this cube,' he says, 'and lived in it, but though I used lighted candles, electrometers, and all other tests of electrical states, I could not find the least influence upon them, or indication

* *Exp. Res.* 1173.

of anything particular given by them, though all the time the outside of the cube was powerfully charged and large sparks and brushes were starting off from every part of its outer surface.'

It appears, therefore, that the most sudden changes of potential produce no physical effects on matter, live or dead, provided these changes take place simultaneously on all the bodies in the field.

If Faraday, instead of raising his cube to a high electric potential, had raised it to a high temperature, the result, as we know, would have been very different.

68.] It appears, therefore, that the analogy between the conduction of heat and electrostatic phenomena has its limits, beyond which we must not attempt to push it. At the time when it was pointed out by Thomson, men of science were already acquainted with the great work of Fourier on the conduction of heat in solid bodies, and their minds were more familiar with the ideas there developed than with those belonging to current electricity, or to the theory of the displacements of a medium.

It is true that Ohm had, in 1827, applied the results obtained by Fourier for heat to the theory of the distribution of electric currents in conductors, but it was long before the practical value of Ohm's work was understood, and till men became familiar with the idea of electric currents in solid conductors, any illustration of electrostatic phenomena drawn from such currents would have served rather to obscure than to enlighten their minds.

69.] When an electric current flows through a solid conductor, the direction of the current at any point is from places of higher to places of lower potential, and its intensity is proportional to the rate at which the potential decreases from point to point of a line drawn in the direction of the current.

We may suppose equipotential surfaces drawn in the conducting medium. The lines of flow of the current are everywhere at right angles to the equipotential surfaces, and the rate of flow is proportional to the number of equipotential surfaces which would be cut by a line of unit length drawn in the direction of the current.

It appears, therefore, that this case of a conducting medium through which an electric current is passing has certain points of analogy with that of a dielectric medium bounded by electrified bodies.

In both the medium is divided into layers by a series of equipotential surfaces. In both there is a system of lines which are everywhere perpendicular to these surfaces. In the one case these lines are called current lines or lines of flow; in the other they are called lines of electric force or electric induction.

An assemblage of such lines drawn from every point of a given line is called a surface of flow. Since the lines of which this surface is formed are everywhere in the direction of the electric current, no part of the current passes through the surface of flow. Such a surface therefore may be regarded as impervious to the current without in any way altering the state of things.

If the line from which the assemblage of lines of flow is drawn is one which returns into itself, which we shall call a *closed* curve, or, more briefly, a *ring*, the surface of flow will have the form of a tube and is called a tube of flow. Any two sections of the same tube of flow correspond to each other in the sense defined in Art. 54, and the quantities of electricity which in the same time flow across these two sections are equal.

Here then we have the analogue of Faraday's law, that the quantities of electricity upon corresponding areas of opposed conducting surfaces are equal and opposite.

Faraday made great use of this analogy between electrostatic phenomena and those of the electric current, or, as he expressed it. between induction in dielectrics and conduction in conductors, and he proved that, in many cases, induction and conduction are associated phenomena. *Exp. Res.* 1320, 1326.

We must remember, however, that the electric current cannot be maintained constant through a conductor which resists its passage except by a continual expenditure of energy, whereas induction in a perfectly insulating dielectric between oppositely electrified conductors may be maintained in it for an indefinitely long time without any expenditure of energy, except that which is required to produce the original electrification. The element of time enters into the question of conduction in a way in which it does not appear in that of induction.

70.] But we may arrive at a more perfect mental representation of induction by comparing it, not with the instantaneous state of a current, but with the small displacements of a medium of invariable density.

Returning to the case of an electric current through a solid conductor, let us suppose that the current, after flowing for a very short time, ceases. If we consider a surface drawn within the solid, then if this surface intersects the tubes of flow, a certain quantity of electricity will have passed from one side of the surface to the other during the time when the current was flowing. This passage of electricity through the surface is called *electric displacement*, and the displacement through a given surface is the quantity of electricity which passes through it. In the case of a continuous current the displacement increases continuously as long as the current is kept up, but if the current lasts for a finite time, the displacement reaches its final value and then remains constant. The lines, surfaces, and tubes of flow of the transient current are also lines, surfaces, and tubes of displacement. The displacements across any two sections of the same tube of displacement are equal. At the beginning of each unit tube of displacement there is a unit of positive electricity, and at the end of the tube there is a unit of negative electricity.

At every point of the medium there is a state of stress consisting of a tension in the direction of the line of displacement through the point and a pressure in all directions at right angles to this line. The numerical value of the tension is equal to that of the pressure, namely, the square of the intensity of the electric force divided by 4π.

71.] By the consideration of the properties of the tubes of induction and the equipotential surfaces we may easily prove several important general theorems in the theory of electricity, the demonstration of which by the older methods is long and difficult. The properties of a tube of induction have already been stated, but for the sake of what follows we may state them again :—

(1) If a tube of induction is cut by an imaginary surface, the quantity of electricity displaced across a section of the tube is the same at whatever part of the tube the section be made.

(2) In every part of the course of a line of electrostatic force it cuts the equipotential surfaces at right angles, and is proceeding from a place of higher to a place of lower potential.

Note. This statement is true only when the distribution of electric force can be completely represented by means of a set of equipotential surfaces. This is always the case when the electricity is in equilibrium, but when there are electric currents, though in some parts of the field it may be possible to draw a set of equipotential surfaces, there are other parts of the field where the distribution of electric force cannot be represented by means of such surfaces. For an electric current is always of the nature of a circuit which returns into itself, and such a circuit cannot in every part of its course be proceeding from places of higher to places of lower potential.

72.] It may be observed that in (1) we have used the words 'tube of induction,' and in (2) the words 'line of electrostatic force.' In a fluid dielectric, such as air, the line of electrostatic force is always in the same direction as the tube of induction, and it may seem pedantic to maintain a distinction between them. There are other cases, however, in which it is very important to remember that a tube of induction is defined with respect to the phenomenon which we have called electric displacement, while a line of force is defined with respect to the electric force. In fluids the electric displacement is always in the direction of the electric force, but there are solid bodies in which this is not the case*, and in which, therefore, the tubes of induction do not coincide in direction with the lines of force.

73.] It follows from (1) that every tube of induction begins at a place where there is a certain quantity of positive electricity and ends at a place where there is an equal quantity of negative electricity, and that, conversely, from any place where there is positive electricity a tube may be drawn, and that wherever there is negative electricity a tube must terminate.

74.] It follows from (2) that the potential at the beginning of a tube is higher than at the end of it. Hence, no tube can return into itself, for in that case the same point would have two different potentials, which is impossible.

75.] From this we may prove that if the potential at every point of a closed surface is the same, and if there is no electrified body within that surface, the potential at any point within the region enclosed within the closed surface is the same as that at the surface.

For if there were any difference of potential between one point and another within this region, there would be lines of force from the places of higher towards the places of lower potential. These lines, as we have seen, cannot return into themselves. Hence they must have their extremities either within the region or without it. Neither extremity of a line of force can be within the region, for there must be positive electrification at the beginning and negative electrification at the end of a line of force, but by our hypothesis there is no electrification within the region. On the other hand, a

* See the experiments of Boltzmann on crystals of sulphur. Vienna Sitzungsb. 9 Jan. 1873.

line of force within the region cannot have its extremities without the region, for in that case it must enter the region at one point of the surface and leave it at another, and therefore by (2) the potential must be higher at the point of entry than at the point of issue, which is contrary to our hypothesis that the potential is the same at every point of the surface.

Hence no line of force can exist within the region, and therefore the potential at any point within the region is the same as that at the surface itself.

76.] It follows from this theorem, that if the closed surface is the internal surface of a hollow conducting vessel, and if no electrified body is within the surface, there is no electrification on the internal surface. For if there were, lines of force would proceed from the electrified parts of the surface into the region within, and we have already proved that there are no lines of force in that region.

We have already proved this by experiment (Art. 20), but we now see that it is a necessary consequence of the properties of lines of force.

Superposition of electric systems.

77.] We have already (Art. 29) given some examples of the superposition of electric effects, but we must now state the principle of superposition more definitely.

If the same system is electrified in three different ways, then if the potential at any point in the third case is the sum of the potentials in the first and second cases, the electrification of any part of the system in the third case will be the sum of the electrifications of the same part in the first and second cases.

By reversing the sign of the electrifications and potentials in one of these cases, we may enunciate the principle with respect to the case in which the potential and the electrification are at every point the excess of what they are in the first case over what they are in the second.

78.] We may now establish a theorem which is of the greatest importance in the theory of electricity.

If the electric field under consideration consist of a finite portion of a dielectric medium, and if at every point of the boundary of this region the potential is given, and if the distribution of electrification within the region be also given, then the potential at any point within the region can have one and only one value consistent with these conditions.

One value at least of the potential must be possible, because the conditions of the theorem are physically possible. Again, if at any point two values of the potential were possible, then by taking the excess of the first value over the second for every point of the system, a third case might be formed in which the potential is everywhere the excess of the first case above the second. At the boundary of the region the potential in the third case is everywhere zero. Within the region the electrification is everywhere zero. Hence, by (Art. 75), at every point within the region the potential in the third case is zero.

There is, therefore, no difference between the distribution of potential in the first case and in the second, or, in other words, the potential at any point within the region can have only one value.

If in any case we can find a distribution of potential which satisfies the given conditions, then by this theorem we are assured that this distribution is the only possible solution of the problem. Hence the importance of this theorem in the theory of electricity.

79.] For instance, let A be an electrified body and let B be one of the equipotential surfaces surrounding the body. Let the potential of the surface B be equal to P. Now let a conducting body be constructed and placed so that its external surface coincides with the closed surface B, and let it be so electrified that its potential is P. Then the conditions of the region outside B are the same as when it was acted on by the body A only. For the potential over the whole bounding surface of the region is P, the same as before, and whatever electrified bodies exist outside of B remain unchanged. Hence the potential at every point outside of B *may*, consistently with the conditions, be the same as before. By our theorem, therefore, the potential at every point outside B *must* be the same, when, instead of the body A, we have a conducting surface B, raised to the potential P.

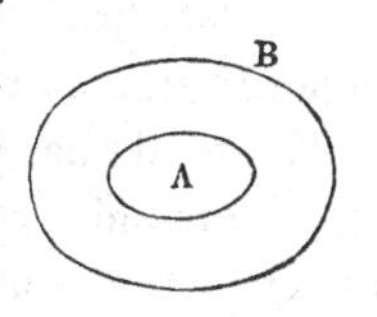

Fig. 18.

80.] The charge of every part of the surface of a conductor is of the same sign as its potential, unless there is another body in the field whose potential is of the same sign but numerically greater.

Let us suppose the potential of the body to be positive; then, if on any part of its surface there is negative electricity, lines of force must terminate on this part of the surface, and these lines of force must begin at some electrified surface whose potential is higher than that of the body. Hence, if there is no other body whose potential is higher than that of the given body, no part of the surface of the given body can be charged with negative electricity.

If an uninsulated conductor is placed in the same field with a charged conductor, the charge on every part of the surface of the uninsulated conductor is of the opposite sign to the charge of the charged conductor.

For since the potential of the uninsulated body is zero, there can be no line of force between it and the walls of the room, or infinite space where the potential is always zero. The line of force which has one end at any point of the surface of this body must therefore have its other end at some point of the charged body, and since the two extremities of a line of force are oppositely electrified, the electrification of the surface of the uninsulated body must be everywhere opposite to the charge of the charged body.

The charged body in this experiment is called the Inductor, and the other body the induced body.

When the induced body is uninsulated, the electricity spread over every part of its surface is, as we have just proved, of the opposite sign to that of the inductor.

The total charge, E_A, of the induced body, which we may call A, may be found by multiplying P_B, the potential of the inductor B, by Q_{AB}, the mutual coefficient of induction between the bodies, which is always a negative quantity.

This electrification induced on an uninsulated body is called by some writers on electricity the Induced Electrification of the First Species. Since the potential of A is already zero, it is manifest that if any part of its surface is touched by a fine wire communicating with the ground there will be no discharge.

Next, let us suppose that the body, A, instead of being uninsulated is insulated, but originally without charge. Under the action of the inductor B part of its surface, on the side next to B, will become electrified oppositely to B; but since the algebraic sum of its electrification is zero, some other part of its surface must be electrified similarly to B.

This electrification, of the same name as that of B, is called by writers on electricity the Induced Electrification of the Second

Species. If a wire connected with the ground be now made to touch any part of the surface of A, electricity of the same name as that of B will be discharged, its amount being equal and opposite to the negative charge (of the first species) which remains on the body A, which is now reduced to potential zero.

In order to obtain a clearer idea of the distribution of electricity on the surface of A under various conditions, let us begin by supposing the potential of A to be zero and that of B to be unity. Let the surface-density at a given point P on the surface of A be $-\sigma_1$, and let the whole charge of A be $-q_{AB}$. The negative sign is prefixed to the symbols of these quantities because the quantities themselves are always negative.

The charge of B in this case is q_B.

Let us next suppose the potential of A to be unity and that of B to be zero, and let the surface density at the point P be now σ_2, and the whole charge on A, q_A.

These quantities are both essentially positive, and q_A is called the *capacity* of A. The value of both is increased on account of the presence of B in the field.

Let us now suppose that the potentials of A and B are P_A and P_B respectively; then the surface density at the point P is

$$\sigma = P_A\sigma_2 - P_B\sigma_1,$$

and the charge of A is

$$E_A = P_A q_A - P_B q_{AB},$$

and that of B is

$$E_B = P_B q_B - P_A q_{AB}. \quad \text{[See Art. 39.]}$$

If A is insulated and without charge $E_A = 0$, which gives

$$P_A = P_B \frac{q_{AB}}{q_A},$$

and the surface density at P is

$$\sigma = \frac{P_B}{q_A}(q_{AB}\sigma_2 - q_A\sigma_1).$$

On a region of the surface of A next to B, σ will be of the opposite sign from P_B; and on a region on the other side from B, σ will be of the same sign with P_B. The boundary between these two regions forms what is called the neutral line, the form and position of which depend on the form and position of A and B.

CHAPTER VI.

PARTICULAR CASES OF ELECTRIFICATION.

81.] A SPHERICAL conductor is electrified and insulated within the concentric spherical internal surface of a conducting vessel.

On account of the perfect symmetry of this system in all directions, it is manifest that the distribution of electricity will be uniform over each of the opposed spherical surfaces, that the lines of force will be in the directions passing through the common centre of the spheres, and that the equipotential surfaces will be spheres having this point for their centre.

If e is the quantity of electricity on the inner sphere and E that on the internal surface of the outer sphere, then by Experiment VIII

$$E = -e. \quad \text{(1)}$$

If r and R are the radii of the spheres, s and S their surfaces, and σ and Σ the surface-densities of the electricity on these surfaces, then by geometry,

$$s = 4\pi r^2, \qquad S = 4\pi R^2, \quad \text{(2)}$$

where π denotes the ratio of the circumference of a circle to its diameter.

The whole charge on either sphere is found by multiplying the surface into the surface-density, or

$$e = s\sigma, \qquad E = S\Sigma. \quad \text{(3)}$$

Hence,

$$\sigma = \frac{e}{4\pi r^2}, \qquad \Sigma = \frac{E}{4\pi R^2}, \quad \text{(4)}$$

and by (1),

$$\Sigma = \frac{-e}{4\pi R^2}. \quad \text{(5)}$$

It appears, therefore, that when the charge, e, of the inner sphere is given, the surface-density, Σ, on the internal surface of the vessel is inversely as the square of the distance of that surface from the centre of the electrified sphere.

Hence by Coulomb's law (Experiment XIII, Art. 47) the electromotive force at the outer spherical surface is inversely as the square of the distance from the centre of the sphere.

This is the law according to which the electric force varies at different distances from a sphere uniformly electrified. The amount of the force is independent of the radius of the inner electrified sphere, and depends only on the whole charge upon it. If we suppose the radius of the inner sphere to become very small till at last the sphere cannot be distinguished from a point, we may imagine the whole charge concentrated at this point, and we may then express our result by saying that the electric action of a uniformly electrified sphere at any point outside the sphere is the same as that of the whole charge of the sphere would be if concentrated at the centre of the sphere.

We must bear in mind, however, that it is physically impossible to charge the small sphere with more than a certain quantity of electricity on each unit of area of its surface. If the surface-density exceed this limit, electricity will fly off in the form of the brush discharge. Hence the idea of an electrified point is a mere mathematical fiction which can never be realised in nature. The imaginary charge concentrated at the centre of the sphere, which produces an effect outside the sphere equivalent to that of the actual distribution of electricity on the surface, is called the *Electrical Image* of that distribution. See Art. 100.

Measurement of Electricity.

82.] We have already described methods of comparing the quantity of electrification on different bodies, but in each case we have only compared one quantity of electricity with another, without determining the absolute value of either. To determine the absolute value of an electric charge we must compare it with some definite quantity of electricity, which we assume as a unit.

The unit of electricity adopted in electrostatics is that quantity of positive or vitreous electricity which, if concentrated in a point, and placed at the unit of distance from an equal charge, also concentrated in a point, would repel it with the unit of mechanical force. The dielectric medium between the two charged points is supposed to be air.

83.] Let us now suppose two bodies, whose dimensions are small compared with the distance between them, to be charged with electricity. Let the charge of the first body be e units of electricity and that of the second e' units, and let the distance between the bodies be r.

Then, since the force varies inversely as the square of the distance, the force with which each unit of electricity in the first body repels each unit of electricity in the second body will be $\frac{1}{r^2}$, and since the number of pairs of units, one in each body, is ee', the whole repulsion between the bodies will be

$$f = \frac{ee'}{r^2}$$

If the charge of the first or the second body is negative we must consider e or e' negative. If the one charge is positive and the other negative, f will be negative, or the force between the bodies will be an attraction instead of a repulsion. If the charges are both positive or both negative, the force between the bodies will be a repulsion.

84.] *Definition.*—The electric or electromotive force at a point is the force which would be experienced by a small body charged with the unit of positive electricity and placed at that point, the electrification of the system being supposed to remain undisturbed by the presence of this unit of electricity.

We shall use the German letter $\mathfrak{E}$ as the symbol of electric force.

85.] Let us now return to the case of a sphere whose radius is r, the external surface of which is uniformly electrified, the surface density of the electrification being σ. As we have already proved, the whole charge of the sphere is

$$e = 4\pi r^2 \sigma.$$

At any point outside the sphere such that the distance from the centre of the sphere is r' the electromotive force, $\mathfrak{E}$, is directed *from* the centre, and its value is

$$\mathfrak{E} = \frac{e}{r'^2}.$$

If the point is close to the surface of the sphere, $r' = r$, and

$$\mathfrak{E} = \frac{e}{r^2} = 4\pi\sigma,$$

or the electric force close to the surface of an electrified sphere is at right angles to the surface and is equal to the surface-density multiplied by 4π.

We have already seen that in all cases the electric force close to the surface of a conductor is at right angles to that surface, and is proportional to the surface-density. We now, by means of this particular case, find that the constant ratio of the electric force to the surface-density is 4π for a uniformly electrified sphere, and therefore this is the ratio for a conductor of any form.

The equation
$$\mathfrak{E} = 4\pi\sigma$$
is the complete expression of the law discovered by Coulomb and referred to in Arts. 47 and 81.

Value of the Potential.

86.] We must next consider the values of the potential at different distances from a small electrified body.

Definition. The electric potential at any point is the work which must be expended in order to bring a body charged with unit of electricity from an infinite distance to that point.

If ψ is the potential at A and ψ that at B, then the work which must be spent by the external agency in overcoming electrical force while carrying a unit of electricity from A to B is $\psi' - \psi$.

The quantity $\psi' - \psi$ would also represent the work which would be done *by the electrical forces* in assisting the transfer of the unit of electricity from B to A if the motion were reversed.

If the force from B to A were constant and equal to $\mathfrak{E}$, then
$$\psi' - \psi = \overline{BA} \cdot \mathfrak{E}.$$
In general, the electric force varies as the body moves from B to A, so that we cannot at once apply this method of finding the difference of potentials. But, by breaking up the path BA into a sufficient number of parts, we may make these parts so small that the electric force may be regarded as uniform during the passage of the body along any one of these parts. We may then ascertain the parts of the work done in each part of the path, and by adding them together, obtain the whole work done during the passage from B to A.

Z C B A
O

Fig. 19.

Let us suppose a unit of electricity placed at O, and let the distances of the points $A, B, C, \ldots Z$ from O be $a, b, c, \ldots z$. The electric force at A is $\frac{1}{a^2}$, at B $\frac{1}{b^2}$, and so on, all in the direction from O to A.

To find the work which must be done in order to bring a unit of electricity from A to B we must multiply the distance AB by the average of the electromotive force at the various points between A and B. The least value of the force is $\frac{1}{a^2}$, and the greatest value is $\frac{1}{b^2}$. Hence the work required is greater than $\frac{AB}{a^2}$ and less than $\frac{AB}{b^2}$. Now AB is $a-b$, and the true value of the work is the excess of the potential at B over that at A. Hence if we now write $A, B, C, \dots Z$ for the potentials at the corresponding points, we may express the work required to bring the unit of electricity from A to B by $B-A$. This quantity therefore is greater than

$$\frac{a-b}{a^2} \text{ or } \left(\frac{1}{b}-\frac{1}{a}\right)\frac{b}{a},$$

but less than

$$\frac{a-b}{b^2} \text{ or } \left(\frac{1}{b}-\frac{1}{a}\right)\frac{a}{b}.$$

We may express this by the double inequality

$$\left(\frac{1}{b}-\frac{1}{a}\right)\frac{b}{a} < B-A < \left(\frac{1}{b}-\frac{1}{a}\right)\frac{a}{b}.$$

Similarly

$$\left(\frac{1}{c}-\frac{1}{b}\right)\frac{c}{b} < C-B < \left(\frac{1}{c}-\frac{1}{b}\right)\frac{b}{c},$$

and so on. The ratios $\frac{a}{b}$, $\frac{b}{c}$, &c., are all greater than unity. Let us suppose that the greatest of these ratios is equal to p. The ratios $\frac{b}{a}$, &c., are the reciprocals of these; they are therefore all less than unity, but none less than $\frac{1}{p}$. Hence

$$\left(\frac{1}{b}-\frac{1}{a}\right)\frac{1}{p} < B-A < \left(\frac{1}{b}-\frac{1}{a}\right)p$$

$$\left(\frac{1}{c}-\frac{1}{b}\right)\frac{1}{p} < C-B < \left(\frac{1}{c}-\frac{1}{b}\right)p$$

....

$$\left(\frac{1}{z}-\frac{1}{y}\right)\frac{1}{p} < Z-Y < \left(\frac{1}{z}-\frac{1}{y}\right)p.$$

Adding these inequalities we find

$$\left(\frac{1}{z}-\frac{1}{a}\right)\frac{1}{p} < Z-A < \left(\frac{1}{z}-\frac{1}{y}\right)p.$$

By increasing the number of points between A and Z and making the intervals between them smaller we may make the greatest ratio, p, as near to unity as we please, and we may therefore assert that, as the line AZ is more and more minutely divided, the quantity p and its reciprocal $\frac{1}{p}$ approach unity as their common limit. In the limit, therefore,

$$Z - A = \frac{1}{z} - \frac{1}{a}.$$

We have thus found the difference between the potentials at A and Z. To determine the actual value of the potential, say at Z, we must refer to the definition of the potential, that it is the work expended in bringing unit of electricity from an infinite distance to the given point. We have therefore in the above expression to suppose the point A removed to an infinite distance from O, in which case the potential A is zero, and the reciprocal of the distance, or $\frac{1}{a}$, is also zero. The equation is therefore reduced to the form

$$Z = \frac{1}{z},$$

or in words, the numerical value of the potential at a given point due to unit of electricity at a given distance is the reciprocal of the number expressing that distance.

If the charge is e, then the potential at a distance z is $\frac{e}{z}$.

The potential due to a number of charges placed at different distances from the given point is found by adding the potentials due to each separate charge, regard being had to the sign of each potential.

87.] Since, as we have seen, the electric force at any point outside a uniformly electrified spherical surface is the same as if the electric charge of the surface had been concentrated at its centre, the potential due to the electrified surface must be, for points outside it,

$$\psi = \frac{e}{r},$$

where e is the whole charge of the surface, and r is the distance of the given point from the centre.

Let a be the radius of the spherical surface, then this expression for the potential is true as long as r is greater than a. At the surface, r is equal to a. The potential at the surface due to its own electrification is therefore

$$\psi_a = \frac{e}{a}$$

[since there can be no discontinuity in the value of the potential between the surface and a point just outside it].

Within the surface there is no electromotive force, and the potential is therefore the same as at the surface for all points within the sphere.

If the potential of the spherical surface is unity, then

$$e = a,$$

or the charge is numerically equal to the radius.

Now the electric capacity of a body in a given field is measured by the charge which raises its potential to unity. Hence the electric capacity of a conducting sphere placed in air at a considerable distance from any other conductor is numerically equal to the radius of the sphere.

If by means of an electrometer we can measure the potential of the sphere, we can ascertain its charge by multiplying this potential by the radius of the sphere. This method of measuring a quantity of electricity was employed by Weber and Kohlrausch in their determination of the ratio of the unit employed in electromagnetic to that employed in electrostatic researches. Since there is no electric force within a uniformly electrified sphere the potential within the sphere is constant and equal to $\frac{e}{a}$.

88.] We are now able to complete the theory of the electrification of two concentric spherical surfaces.

Let a spherical conductor of radius a be insulated within a hollow conducting vessel, the internal surface of which is a sphere of radius b concentric with the inner sphere. Let the charge on the inner sphere be e, then, as we have already seen, the charge on the interior surface of the vessel will be $-e$. At any point outside both spherical surfaces and distant r from the centre the electric potential due to the inner sphere will be $\frac{e}{r}$, and that due to the outer sphere will be $\frac{-e}{r}$. Since these two quantities are numerically equal, but of opposite sign, they destroy each other, and the potential at every point for which r is greater than b is zero.

Between the two spherical surfaces, at a point distant r from the centre, the potential due to the inner sphere is $\frac{e}{r}$, and that due to the outer sphere is $\frac{-e}{b}$. Hence the whole potential in this intermediate space is $e\left(\frac{1}{r}-\frac{1}{b}\right)$.

At the surface of the inner sphere $r = a$ so that the potential of the inner sphere is $e\left(\frac{1}{a}-\frac{1}{b}\right)$.

The potential at all points within the inner sphere is uniform and equal to $e\left(\frac{1}{a}-\frac{1}{b}\right)$.

The capacity of the inner sphere is numerically equal to the value of e when the potential is made equal to unity. In this case

$$e = \frac{1}{\frac{1}{a}-\frac{1}{b}} = \frac{ab}{b-a},$$

or, the capacity of a sphere insulated within a concentric spherical surface is a fourth proportional to the distances $(b-a)$ between the surfaces and the radii $(a, b,)$ of the surfaces.

By diminishing the interval, $b-a$, between the surfaces, the capacity of the system may be made very great without making use of very large spheres.

This example may serve to illustrate the principle of the Leyden jar, which consists of two metallic surfaces separated by insulating material. The smaller the distance between the surfaces and the greater the area of the surfaces, the greater the capacity of the jar.

Hence, if an electrical machine which can charge a body up to a given potential is employed to charge a Leyden jar, one surface of which is connected with the earth, it will, if worked long enough, communicate a much greater charge to the jar than it would to a very large insulated body placed at a great distance from any other conductor.

The capacity of the jar, however, depends on the nature of the dielectric which is between the two metallic surfaces as well as on its thickness and area. See Art. 131 et sqq.

Two Parallel Planes.

89. Another simple case of electrification is that in which the electrodes are two parallel plane surfaces at a distance c. We shall suppose the dimensions of these surfaces to be very great compared with the distance between them, and we shall consider the electrical action only in that part of the space between the planes whose distance from the edges of the plates is many times greater than c.

Let A be the potential of the upper plane in the figure, and B that of the lower plane. Then the electric force at any point P between the planes, and not near the edge of either plane, is $\frac{A-B}{c}$, acting from A to B. The electric density on the upper plane is found by Coulomb's Law by dividing this quantity by 4π. If σ be the surface density

Fig. 20.

$$\sigma = \frac{A-B}{4\pi c}. \tag{1}$$

The surface density on the plane B is equal to this in magnitude but opposite in sign.

Let us now consider the quantity of electricity on an area S, which we may suppose cut out from the upper plane by an imaginary closed curve. Multiplying S into σ, we find

$$e = \frac{A-B}{4\pi c} S. \tag{2}$$

The quantity of electricity on an equal area of the plane B taken exactly opposite to S will be $-e$. The energy of the electrification of these two portions of electricity is, by Art. 31,

$$Q = \tfrac{1}{2}\{Ae + B(-e)\} = \tfrac{1}{2}(A-B)e. \tag{3}$$

Expressing this in terms of e it becomes

$$Q = \frac{2\pi}{S} e^2 c. \tag{4}$$

If c, the distance between the surfaces, be made to increase to c' the charges of the surfaces remaining the same, the energy will become

$$Q' = \frac{2\pi}{S} e^2 c'. \tag{5}$$

The augmentation of the potential energy is

$$Q'-Q = \frac{2\pi}{S}e^2(c'-c), \tag{6}$$

and this is the work done by external agency in pulling the planes asunder against the electric attraction.

If F is the electric attraction between the two areas S,

$$F(c'-c) = \frac{2\pi}{S}e^2(c'-c), \tag{7}$$

or
$$F = \frac{2\pi}{S}e^2. \tag{8}$$

90.] This result gives us the best experimental method of measuring the quantity of electricity on the area S, for by this equation

$$e = \sqrt{\frac{FS}{2\pi}}. \tag{9}$$

In this expression F is the force of attraction on the area S determined in dynamical measure from observation of its effects. S is the area of the surface and π is the ratio of the circumference of a circle to its diameter.

The difference between the potentials, A and B, of the two planes is easily found in terms of e by means of equation (2), thus,

$$A-B = 4\pi c\frac{e}{S} = c\sqrt{\frac{8\pi F}{S}}. \tag{10}$$

91.] In Sir William Thomson's attracted disk electrometers a disk is so arranged that when in its proper position the surface of the disk forms part of a much larger plane surface extending for a considerable distance on all sides of the disk. The part of the surface outside the moveable disk is called the Guard Ring and the surface of the disk and guard ring together may be considered as the surface of a large disk, part of which, near its centre, is moveable. Opposite this disk is placed another disk having its surface parallel to the first disk and much larger than the moveable disk. The electrification of the moveable disk is then the same as that of a small portion of one of the large opposed planes taken at a considerable distance from the edge of the plane, and only very small corrections are needed to make the formulæ already given apply to the case of the moveable disk.

The distribution of electrification and of electric force near the edges of the large disks is by no means so simple. It is calculated in Art. 202 of my larger Treatise, and the lines of force and equipotential surfaces are shown in Plate V at the end of this book.

92.] The direct problem of electrostatics—the problem which the circumstances of every electrostatic experiment present to us—may be stated as follows.

A system of insulated conductors is given in form and position, and the electric charge of each conductor is given, required the distribution of electricity on each conductor and the electric potential at any point of the field.

The mathematical difficulties of the solution of this problem have been overcome hitherto only in a small number of cases, and it is only by a study of what we may call the inverse problem that the results we possess have been obtained.

In the inverse problem, a possible distribution of potential being given, it is required to find the forms, positions and charges of a system of conductors which shall be consistent with this distribution of potential.

Any number of solutions of this latter problem may be obtained by taking, instead of the electrified bodies of the original distribution, any set of equipotential surfaces surrounding them, and supposing these surfaces to be the surfaces of conductors, the charge of each conductor being equal to the sum of the charges of all the bodies of the original distribution which it encloses.

Every electrical problem of which we know the solution has been constructed by an inverse process of this kind. It is therefore of great importance to the electrician that he should know what results have been obtained in this way, since the only method by which he can expect to solve a new problem is by reducing it to one of the cases in which a similar problem has been constructed by the inverse process.

This historical knowledge of results can be turned to account in two ways. If we are required to devise an instrument for making electrical measurements with the greatest accuracy, we may select these forms for the electrified surfaces which correspond to cases of which we know the accurate solution. If, on the other hand, we are required to estimate what will be the electrification of bodies whose forms are given, we may begin with some case in which one of the equipotential surfaces takes a form somewhat resembling the given form, and then by a tentative method we may modify the problem till it more nearly corresponds to the given case. This method is evidently very imperfect, considered from a mathematical point of view, but it is the only one we have, and if we are not allowed to choose our conditions, we can make only an approximate calculation of the electrification. It appears, therefore, that what we want is a knowledge of the forms of equipotential surfaces and lines of induction in as many different cases as we can collect together and remember. In certain classes of cases, such as those relating to spheres, we may proceed by mathematical methods. In other cases we cannot afford to despise the humbler method of actually drawing tentative figures on paper, and selecting that which appears least unlike the figure we require.

This latter method, I think, may be of some use, even in cases in which the exact solution has been obtained, for I find that an eye knowledge of the forms of the equipotential surfaces often leads to a right selection of a mathematical method of solution.

I have therefore drawn several diagrams of systems of equipotential surfaces and lines of force, so that the student may make himself familiar with the forms of the lines.

93.] In the first plate at the end of this volume we have the equipotential surfaces surrounding two points electrified with quantities of electricity of the same kind and in the ratio of 20 to 5.

Here each point is surrounded by a system of equipotential surfaces which become more nearly spheres as they become smaller, but none of them are accurately spheres. If two of these surfaces, one surrounding each sphere, be taken to represent the surfaces of two conducting bodies, nearly but not quite spherical, and if these bodies be charged with the same kind of electricity, the charges being as 4 to 1, then the diagram will represent the equipotential surfaces, provided we expunge all those which are drawn inside the two bodies. It appears from the diagram that the action between the bodies will be the same as that between two points having the same charges, these points being not exactly in the middle of the axis of each body, but somewhat more remote than the middle point from the other body.

The same diagram enables us to see what will be the distribution of electricity on one of the oval figures, larger at one end than the other, which surround both centres. Such a body, if electrified with a charge 25 and free from external influence, will have the surface-density greatest at the small end, less at the large end, and least in a circle somewhat nearer the smaller than the larger end.

There is one equipotential surface, indicated by a dotted line, which consists of two lobes meeting at the conical point P. That point is a point of equilibrium, and the surface-density on a body of the form of this surface would be zero at this point.

The lines of force in this case form two distinct systems, divided from one another by a surface of the sixth degree, indicated by a dotted line, passing through the point of equilibrium, and somewhat resembling one sheet of the hyperboloid of two sheets.

This diagram may also be taken to represent the lines of force and equipotential surfaces belonging to two spheres of gravitating matter whose masses are as 4 to 1.

94.] In the second Plate we have again two points whose charges are as 4 to 1, but the one positive and the other negative. In this case one of the equipotential surfaces, that, namely, corresponding to potential zero, is a sphere. It is marked in the diagram by the dotted circle Q. The importance of this spherical surface will be seen when we come to the theory of Electrical Images.

We may see from this diagram that if two round bodies are charged with opposite kinds of electricity they will attract each other as much as two points having the same charges but placed somewhat nearer together than the middle points of the round bodies.

Here, again, one of the equipotential surfaces, indicated by a dotted line, has two lobes, an inner one surrounding the point whose charge is 5 and an outer one surrounding both bodies, the two lobes meeting in a conical point P which is a point of equilibrium.

If the surface of a conductor is of the form of the outer lobe, a roundish body having, like an apple, a conical dimple at one end of its axis, then, if this conductor be electrified, we shall be able to determine the superficial density at any point. That at the bottom of the dimple will be zero.

Surrounding this surface we have others having a rounded dimple which flattens and finally disappears in the equipotential surface passing through the point marked M.

The lines of force in this diagram form two systems divided by a surface which passes through the point of equilibrium.

If we consider points on the axis on the further side of the point B, we find that the resultant force diminishes to the double point P, where it vanishes. It then changes sign, and reaches a maximum at M, after which it continually diminishes.

This maximum, however, is only a maximum relatively to other points on the axis, for if we draw a surface perpendicular to the axis, M is a point of minimum force relatively to neighbouring points on that surface.

95.] Plate III represents the equipotential surfaces and lines of force due to an electrified point whose charge is 10 placed at A, and surrounded by a field of force, which, before the introduction of the electrified point, was uniform in direction and magnitude at every part. In this case, those lines of force which belong to A are contained within a surface of revolution which has an asymptotic cylinder, having its axis parallel to the undisturbed lines of force.

The equipotential surfaces have each of them an asymptotic plane. One of them, indicated by a dotted line, has a conical point and a lobe surrounding the point A. Those below this surface have one sheet with a depression near the axis. Those above have a closed portion surrounding A and a separate sheet with a slight depression near the axis.

If we take one of the surfaces below A as the surface of a conductor, and another a long way below A as the surface of another conductor at a different potential, the system of lines and surfaces between the two conductors will indicate the distribution of electric force. If the lower conductor is very far from A its surface will be very nearly plane, so that we have here the solution of the distribution of electricity on two surfaces, both of them nearly plane and parallel to each other, except that the upper one has a protuberance near its middle point, which is more or less prominent according to the particular equipotential line we choose for the surface.

96.] Plate IV represents the equipotential surfaces and lines of force due to three electrified points A, B and C, the charge of A being 15 units of positive electricity, that of B 12 units of negative electricity, and that of C 20 units of positive electricity. These points are placed in one straight line, so that

$$AB = 9, \quad BC = 16, \quad AC = 25.$$

In this case, the surface for which the potential is unity consists of two spheres whose centres are A and C and their radii 15 and 20. These spheres intersect in the circle which cuts the plane of the paper in D and D', so that B is the centre of this circle and its radius is 12. This circle is an example of a line of equilibrium, for the resultant force vanishes at every point of this line.

If we suppose the sphere whose centre is A to be a conductor with a charge of 3 units of positive electricity, and placed under the influence of 20 units of positive electricity at C, the state of the case will be represented by the diagram if we leave out all the lines within the sphere A. The part of this spherical surface within the small circle DD' will be negatively electrified by the influence of C. All the rest of the sphere will be positively electrified, and the small circle DD' itself will be a line of no electrification.

We may also consider the diagram to represent the electrification of the sphere whose centre is C, charged with 8 units of positive electricity, and influenced by 15 units of positive electricity placed at A.

The diagram may also be taken to represent the case of a conductor whose surface consists of the larger segments of the two spheres meeting in DD', charged with 23 units of positive electricity.

97.] I am anxious that these diagrams should be studied as illustrations of the language of Faraday in speaking of 'lines of force,' the 'forces of an electrified body,' &c.

In strict mathematical language the word Force is used to signify the supposed cause of the tendency which a material body is found to have towards alteration in its state of rest or motion. It is indifferent whether we speak of this observed tendency or of its immediate cause, since the cause is simply inferred from the effect, and has no other evidence to support it.

Since, however, we are ourselves in the practice of directing the motion of our own bodies, and of moving other things in this way, we have acquired a copious store of remembered sensations relating to these actions, and therefore our ideas of force are connected in our minds with ideas of conscious power, of exertion, and of fatigue, and of overcoming or yielding to pressure. These ideas, which give a colouring and vividness to the purely abstract idea of force, do not in mathematically trained minds lead to any practical error.

But in the vulgar language of the time when dynamical science was unknown, all the words relating to exertion, such as force, energy, power, &c., were confounded with each other, though some of the schoolmen endeavoured to introduce a greater precision into their language.

The cultivation and popularization of correct dynamical ideas since the time of Galileo and Newton has effected an immense change in the language and ideas of common life, but it is only within recent times, and in consequence of the increasing importance of machinery, that the ideas of force, energy, and power have become accurately distinguished from each other. Very few, however, even of scientific men, are careful to observe these distinctions; hence we often hear of the force of a cannon-ball when either its energy or its momentum is meant, and of the force of an electrified body when the quantity of its electrification is meant.

Now the quantity of electricity in a body is measured, according to Faraday's ideas, by the *number* of lines of force, or rather of induction, which proceed from it. These lines of force must all terminate somewhere, either on bodies in the neighbourhood, or on the walls and roof of the room, or on the earth, or on the heavenly bodies, and wherever they terminate there is a quantity of electricity exactly equal and opposite to that on the part of the body from which they proceeded. By examining the diagrams this will be seen to be the case. There is therefore no contradiction between Faraday's views and the mathematical result of the old theory, but, on the contrary, the idea of lines of force throws great light on these results, and seems to afford the means of rising by a continuous process from the somewhat rigid conceptions of the old theory to notions which may be capable of greater expansion, so as to provide room for the increase of our knowledge by further researches.

98.] These diagrams are constructed in the following manner:—

First, take the case of a single centre of force, a small electrified body with a charge E. The potential at a distance r is $V = \frac{E}{r}$; hence, if we make $r = \frac{E}{V}$, we shall find r, the radius of the sphere for which the potential is V. If we now give to V the values 1, 2, 3, &c., and draw the corresponding spheres, we shall obtain a series of equipotential surfaces, the potentials corresponding to which are measured by the natural numbers. The sections of these spheres by a plane passing through their common centre will be circles, which we may mark with the number denoting the potential of each. These are indicated by the dotted circles on the right hand of Fig. 21.

If there be another centre of force, we may in the same way draw the equipotential surfaces belonging to it, and if we now wish to find the form of the equipotential surfaces due to both centres together, we must remember that if V_1 be the potential due to one centre, and V_2 that due to the other, the potential due to both will be $V_1 + V_2 = V$. Hence, since at every intersection of the equipotential surfaces belonging to the two series we know both V_1 and V_2, we also know the value of V. If therefore we draw a surface which passes through all those intersections for which the value of V is the same, this surface will coincide with a true equipotential surface at all these intersections, and if the original systems of surfaces be drawn sufficiently close, the new surface may be drawn with any required degree of accuracy. The equipotential surfaces due to two points whose charges are equal and opposite are represented by the continuous lines on the right hand side of Fig. 21.

This method may be applied to the drawing of any system of equipotential surfaces when the potential is the sum of two potentials, for which we have already drawn the equipotential surfaces.

The lines of force due to a single centre of force are straight lines radiating from that centre. If we wish to indicate by these lines the intensity as well as the direction of the force at any point, we must draw them so that they mark out on the equipotential surfaces portions over which the surface-integral of induction has definite values. The best way of doing this is to suppose our plane figure to be the section of a figure in space formed by the revolution of the plane figure about an axis passing through the centre of force. Any straight line radiating from the centre and making an angle θ with the axis will then trace out a cone, and the surface-integral of the induction through that part of any surface which is cut off by this cone on the side next the positive direction of the axis, is $2\pi E(1-\cos\theta)$.

If we further suppose this surface to be bounded by its intersection with two planes passing through the axis, and inclined at the angle whose arc is equal to half the radius, then the induction through the surface so bounded is

$$E(1-\cos\theta) = 2\Psi, \text{ say};$$

$$\text{and} \quad \theta = \cos^{-1}\left(1 - 2\frac{\Psi}{E}\right).$$

If we now give to Ψ a series of values 1, 2, 3 ... E, we shall find a corresponding series of values of θ, and if E be an integer, the number of corresponding lines of force, including the axis, will be equal to E.

We have therefore a method of drawing lines of force so that the charge of any centre is indicated by the number of lines which converge to it, and the induction through any surface cut off in the

To face P. 78.

Fig: 21.

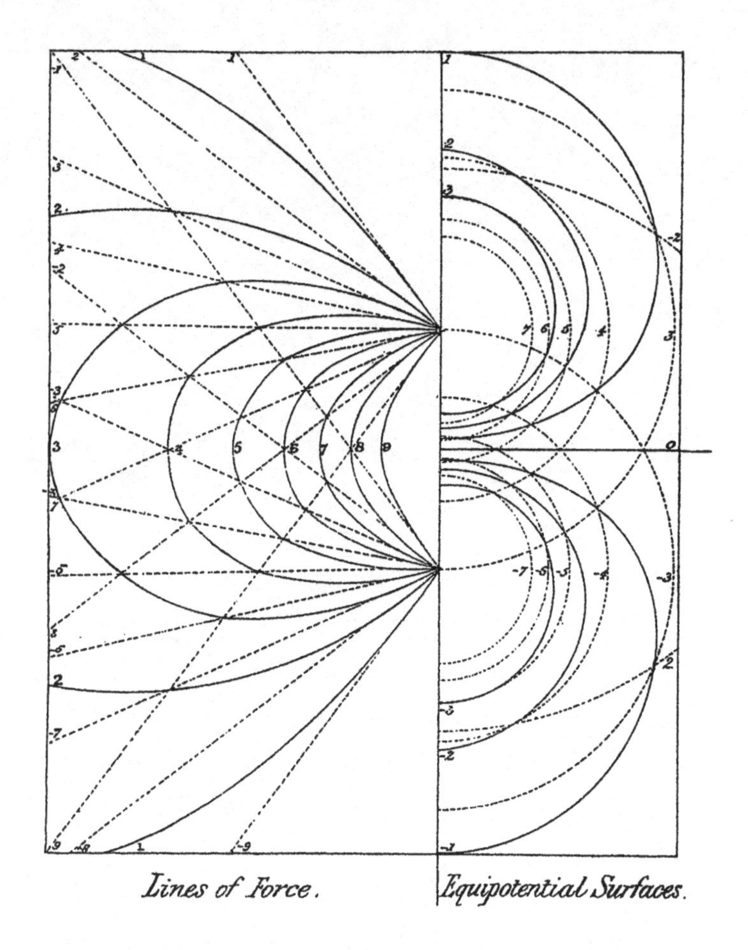

Method of drawing
Lines of Force and Equipotential Surfaces.

For the Delegates of the Clarendon Press.

way described is measured by the number of lines of force which pass through it. The dotted straight lines on the left hand side of Fig. 21 represent the lines of force due to each of two electrified points whose charges are 10 and -10 respectively.

If there are two centres of force on the axis of the figure we may draw the lines of force for each axis corresponding to values of Ψ_1 and Ψ_2, and then, by drawing lines through the consecutive intersections of these lines, for which the value of $\Psi_1+\Psi_2$ is the same, we may find the lines of force due to both centres, and in the same way we may combine any two systems of lines of force which are symmetrically situated about the same axis. The continuous curves on the left hand side of Fig. 21 represent the lines of force due to the two electrified points acting at once.

After the equipotential surfaces and lines of force have been constructed by this method the accuracy of the drawing may be tested by observing whether the two systems of lines are everywhere orthogonal, and whether the distance between consecutive equipotential surfaces is to the distance between consecutive lines of force as half the distance from the axis is to the assumed unit of length.

In the case of any such system of finite dimensions the line of force whose index number is Ψ has an asymptote which passes through the centre of gravity of the system, and is inclined to the axis at an angle whose cosine is $1-2\dfrac{\Psi}{E}$, where E is the total electrification of the system, provided Ψ is less than E. Lines of force whose index is greater than E are finite lines.

The lines of force corresponding to a field of uniform force parallel to the axis are lines parallel to the axis, the distances from the axis being the square roots of an arithmetical series.

CHAPTER VII.

THEORY OF ELECTRICAL IMAGES.

99.] The calculation of the distribution of electrification on the surface of a conductor when electrified bodies are placed near it is in general an operation beyond the powers of existing mathematical methods.

When the conductor is a sphere, and when the distribution of electricity on external bodies is given, a solution, depending on an infinite series was obtained by Poisson. This solution agrees with that which was afterwards given in a far simpler form by Sir W. Thomson, and which is the foundation of his method of Electric Images.

By this method he has solved problems in electricity which have never been attempted by any other method, and which, even after the solution has been pointed out, no other method seems capable of attacking. This method has the great advantage of being intelligible by the aid of the most elementary mathematical reasoning, especially when it is considered in connection with the diagrams of equipotential surfaces described in Arts. 93–96.

100.] The idea of an image is most easily acquired by considering the optical phenomena on account of which the term image was first introduced into science.

We are accustomed to make use of the visual impressions we receive through our eyes in order to ascertain the positions of distant objects. We are doing this all day long in a manner sufficiently accurate for ordinary purposes. Surveyors and astronomers by means of artificial instruments and mathematical deductions do the same thing with greater exactness. In whatever way, however, we make our deductions we find that they are consistent with the hypothesis that an object exists in a certain position in space, from which it emits light which travels to our eyes or to our instruments in straight lines.

But if we stand in front of a plane mirror and make observations on the apparent direction of the objects reflected therein, we find that these observations are consistent with the hypothesis that there is no mirror, but that certain objects exist in the region beyond the plane of the mirror. These hypothetical objects are geometrically related to certain real objects in front of the plane of the mirror, and they are called the *images* of these objects.

We are not provided with a special sense for enabling us to ascertain the presence and the position of distant bodies by means of their electrical effects, but we have instrumental methods by which the distribution of potential and of electric force in any part of the field may be ascertained, and from these data we obtain a certain amount of evidence as to the position and electrification of the distant body.

If an astronomer, for instance, could ascertain the direction and magnitude of the force of gravitation at any desired point in the heavenly spaces, he could deduce the positions and masses of the bodies to which the force is due. When Adams and Leverrier discovered the hitherto unknown planet Neptune, they did so by ascertaining the direction and magnitude of the gravitating force due to the unseen planet at certain points of space. In the electrical problem we employed an electrified pith ball, which we moved about in the field at pleasure. The astronomers employed for a similar purpose the planet Uranus, over which, indeed, they had no control, but which moved of itself into such positions that the alterations of the elements of its orbit served to indicate the position of the unknown disturbing planet.

101.] In one of the electrified systems which we have already investigated, that of a spherical conductor A within a concentric spherical conducting vessel B, we have one of the simplest cases of the principle of electric images.

The electric field is in this case the region which lies between the two concentric spherical surfaces. The electric force at any point P within this region is in the direction of the radius OP and numerically equal to the charge of the inner sphere, A, divided by the square of the distance, OP, of the point from the common centre. It is evident, therefore, that the force within this region will be the same if we substitute for the electrified spherical surfaces, A and B, any other two concentric spherical surfaces, C and D, one of them, C, lying within the smaller sphere, A, and the other, D, lying outside of B, the charge of C being equal to that of A in the former case. The electric phenomena in the region between A and B are therefore the same as before, the only difference between the cases is that in the region between A and C and also in the region between B and D we now find electric forces acting according to the same law as in the region between A and B, whereas when the region was bounded by the conducting surfaces A and B there was no electrical force whatever in the regions beyond these surfaces. We may even, for mathematical purposes, suppose the inner sphere C to be reduced to a physical point at O, and the outer sphere D to expand to an infinite size, and thus we assimilate the electric action in the region between A and B to that due to an electrified point at O placed in an infinite region.

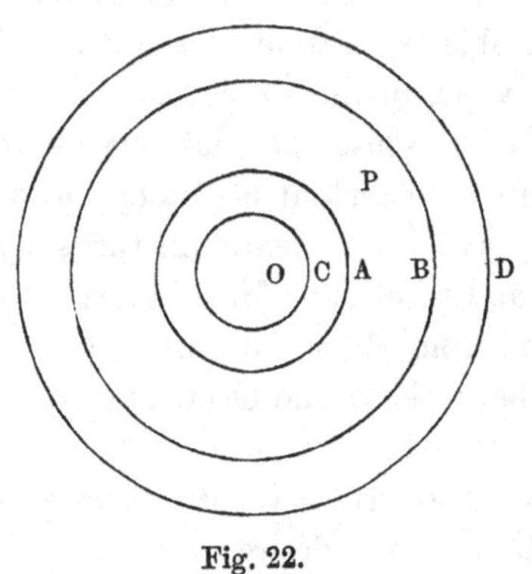

Fig. 22.

It appears, therefore, that when a spherical surface is uniformly electrified, the electric phenomena in the region outside the sphere are exactly the same as if the spherical surface had been removed, and a very small body placed at the centre of the sphere, having the same electric charge as the sphere.

This is a simple instance in which the phenomena in a certain region are consistent with a false hypothesis as to what exists beyond that region. The action of a uniformly electrified spherical surface in the region outside that surface is such that the phenomena may be attributed to an imaginary electrified point at the centre of the sphere.

The potential, ψ, of a sphere of radius a, placed in infinite space and charged with a quantity e of electricity, is $\frac{e}{a}$. Hence if ψ is the potential of the sphere, the imaginary charge at its centre is ψa.

102.] Now let us calculate the potential at a point P in a spherical surface whose centre is C and radius $\overline{CP}$, due to two electrified points A and B in the same radius produced, and such that the product of their distances from the centre is equal to the square of the radius. Points thus related to one another are called *inverse* points with respect to the sphere.

Let $a = \overline{CP}$ be the radius of the sphere. Let $\overline{CA} = ma$, then $\overline{CB}$ will be $\dfrac{a}{m}$.

Also the triangle APC is similar to PCB, and

$$\overline{AP} : \overline{PB} :: \overline{AC} : \overline{PC},$$

or

$$\overline{AP} = m\overline{BP}. \quad \text{See Euclid vi. prop. E.}$$

Now let a charge of electricity equal to e be placed at A and a charge $e' = -\dfrac{e}{m}$ of the opposite kind be placed at B. The potential due to these charges at P will be

$$V = \frac{e}{\overline{AP}} + \frac{e'}{\overline{BP}},$$

$$= \frac{e}{m\overline{BP}} - \frac{e}{m\overline{BP}},$$

$$= 0;$$

or the potential due to the charges at A and B at any point P of the spherical surface is zero.

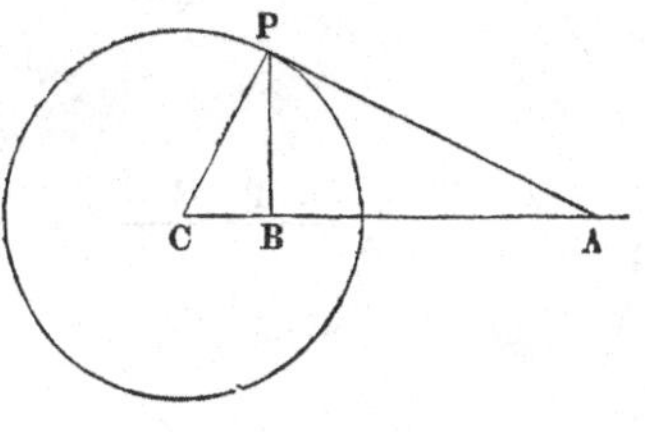

Fig. 23.

We may now suppose the spherical surface to be a thin shell of metal. Its potential is already zero at every point, so that if we connect it by a fine wire with the earth there will be no alteration of its potential, and therefore the potential at every point, whether within or without the surface, will remain unaltered, and will be that due to the two electrified points A and B.

If we now keep the metallic shell in connection with the earth and remove the electrified point B, the potential at every point within the sphere will become zero, but outside it will remain as before. For the surface of the sphere still remains of the same potential, and no change has been made in the distribution of electrified bodies in the region outside the sphere.

Hence, if an electrified point A be placed outside a spherical conductor which is at potential zero, the electrical action at all points outside the sphere will be equivalent to that due to the point A together with another point, B, within the sphere, which is the inverse point to A, and whose charge is to that of A as -1 is to m. The point B with its imaginary charge is called the *electric image* of A.

In the same way by removing A and retaining B, we may shew that if an electrified point B be placed inside a hollow conductor having its inner surface spherical, the electrical action within the hollow is equivalent to that of the point B, together with an imaginary point, A, outside the sphere, whose charge is to that of B as m is to -1.

If the sphere, instead of being in connection with the earth, and therefore at potential zero, is at potential ψ, the electrical effects outside the sphere will be the same as if, in addition to the image of the electrified point, another imaginary charge equal to ψa were placed at the centre of the sphere.

Within the sphere the potential will simply be increased by ψ.

103.] As an example of the method of electric images let us calculate the electric state of two spheres whose radii are a and b respectively, and whose potentials are P_a and P_b, the distance between their centres being c. We shall suppose b to be small compared with c.

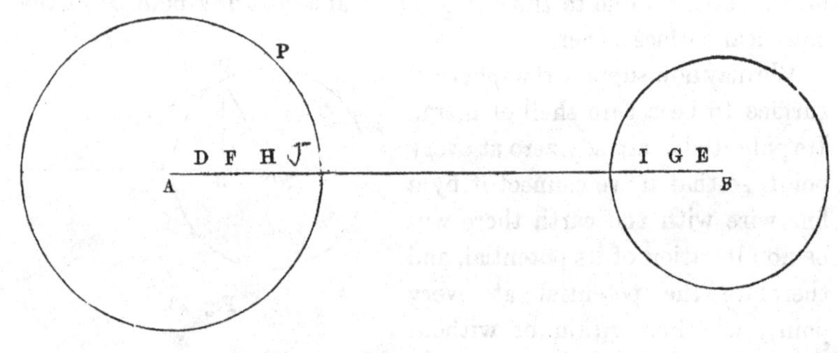

Fig. 24.

We may consider the actual electrical effects at any point outside the two spheres as due to a series of electric images.

In the first place, since the potential of the sphere A is P_a we must place an image at the centre A with a charge aP_a.

Similarly at B, the centre of the other sphere, we must place a charge bP_b.

Each of these images will have an image of the second order in the other sphere. The image of B in the sphere a will be at D, where

$$AD = \frac{a^2}{c}, \text{ and the charge } D = -\frac{a}{c}\cdot bP_b.$$

The image of A in the sphere b will be at E, where

$$BE = \frac{b^2}{c}, \text{ and the charge } E = -\frac{b}{c}\cdot aP_a.$$

Each of these will have an image of the third order. That of E in a will be at F, where

$$AF = \frac{a^2}{AE} = \frac{a^2 c}{c^2 - b^2}, \text{ and } F = \frac{a^2 b}{c^2 - b^2} P_a.$$

That of D in b will be at G, where

$$BG = \frac{b^2}{DB} = \frac{b^2 c}{c^2 - a^2}, \text{ and } G = \frac{ab^2}{c^2 - a^2} P_b.$$

The images of the fourth order will be,

of G in a at H, where

$$AH = \frac{a^2}{AG} = \frac{a^2 (c^2 - a^2)}{c(c^2 - a^2 - b^2)} \text{ and } H = \frac{a^2 b^2}{c(c^2 - a^2 - b^2)} P_b,$$

of F in b at I, where

$$BI = \frac{b^2}{FB} = \frac{b^2 (c^2 - b^2)}{c(c^2 - a^2 - b^2)} \text{ and } I = \frac{a^2 b^2}{c(c^2 - a^2 - b^2)} P_a.$$

We might go on with a series of images for ever, but if b is small compared with c, the images will rapidly become smaller and may be neglected after the fourth order.

If we now write

$$q_{aa} = a + \frac{a^2 b}{c^2 - b^2} + \text{\&c.},$$

$$q_{ab} = -\frac{ab}{c} - \frac{a^2 b^2}{c(c^2 - a^2 - b^2)} - \text{\&c.},$$

$$q_{bb} = b + \frac{ab^2}{c^2 - a^2} + \text{\&c.},$$

the whole charge of the sphere a will be

$$E_a = q_{aa} P_a + q_{ab} P_b,$$

and that of the sphere b will be

$$E_b = q_{ab} P_a + q_{bb} P_b.$$

104.] From these results we may calculate the potentials of the two spheres when their charges are given, and if we neglect terms involving b^3 we find

$$P_a = \frac{1}{a} E_a + \frac{1}{c} E_b,$$

$$P_b = \frac{1}{c} E_a + \left\{ \frac{1}{b} - \frac{a^3}{c^2 (c^2 - a^2)} \right\} E_b.$$

The electric energy of the system is

$$\frac{1}{2}\left(E_a P_a + E_b P_b\right) = \frac{1}{2}\frac{1}{a} E_a^2 + \frac{1}{c} E_a E_b + \frac{1}{2}\left\{ \frac{1}{b} - \frac{a^3}{c^2 (c^2 - a^2)} \right\} E_b^2.$$

The repulsion, R, between the two spheres is measured by the rate at which the energy diminishes as c increases; therefore,

$$R = \frac{E_b}{c^2}\left\{E_a - E_b\frac{a^3(2c^2-a^2)}{c(c^2-a^2)^2}\right\}.$$

In order that the force may be repulsive it is necessary that the charges of the spheres should be of the same sign, and

$$E_a \text{ must be greater than } E_b\frac{a^3(2c^2-a^2)}{c(c^2-a^2)^2}.$$

Hence the force is always attractive,

1. When either sphere is uninsulated;
2. When either sphere has no charge;
3. When the spheres are very nearly in contact, if their potentials are different.

When the potentials of the two spheres are equal the force is always repulsive.

105.] To determine the electric force at any point just outside of the surface of a conducting sphere connected with the earth arising from the presence of an electrified point A outside the sphere.

The electrical conditions at all points outside the sphere are equivalent, as we have seen, to those due to the point A together with its image at B. If e is the charge of the point A (Fig. 23), the force due to it at P is $\frac{e}{AP^2}$ in the direction AP. Resolving this force in a direction parallel to AC and along the radius, its components are $\frac{e}{AP^3}AC$ in the direction parallel to AC and $\frac{e}{AP^3}CP$ in the direction CP. The charge of the image of A at B is $-e\frac{CP}{CA}$, and the force due to the image at P is $e\frac{CP}{CA}\cdot\frac{1}{BP^2}$ in the direction PB. Resolving this force in the same direction as the other, its components are

$$e\frac{CP}{CA}\cdot\frac{CB}{BP^3} \text{ in a direction parallel to } CA, \text{ and}$$

$$e\frac{CP^2}{CA\,.\,BP^3} \text{ in the direction } PC.$$

If a is the radius of the sphere and if $CA = f = ma$ and $AP = r$, then $CB = \frac{1}{m}a$ and $BP = \frac{1}{m}r$; and if e is the charge of the point A, the charge of its image at B is $-\frac{1}{m}e$.

The force at P due to the charge e at A is $\frac{e}{r^2}$ in the direction AP.

Resolving this force in the direction of the radius and a direction parallel to AC, its components are

$$\frac{e}{r^2}\cdot\frac{ma}{r} \text{ in the direction } AC, \text{ and}$$

$$\frac{e}{r^2}\cdot\frac{a}{r} \text{ in the direction } CP.$$

The force at P due to the image $-\frac{1}{m}e$ at B is $\frac{1}{m}e\frac{1}{BP^2}$ or $e\frac{m}{r^2}$ in the direction PB. Resolving this in the same directions as the other force, its components are

$$e\frac{m}{r^2}\frac{BC}{BP} = \frac{ema}{r^3} \text{ in the direction } CA, \text{ and}$$

$$e\frac{m\,.\,CP}{r^2\,BP} \text{ or } \frac{em^2a}{r^3} \text{ in the direction } PC.$$

The components in the direction parallel to AC are equal but in opposite directions. The resultant force is therefore in the direction of the radius, which confirms what we have already proved, that the sphere is an equipotential surface to which the resultant force is everywhere normal. The resultant force is therefore in the direction PC, and is equal to $\frac{ea}{r^3}(m^2-1)$ in the direction PC, that is to say, towards the centre of the sphere.

From this we may ascertain the surface density of the electrification at any point of the sphere, for, by Coulomb's law, if σ is the surface density,

$4\pi\sigma = R$, where R is the resultant force *acting outwards*.

Hence, as the resultant force in this case acts inwards, the surface density is everywhere negative, and is

$$\sigma = -\frac{1}{4\pi}\frac{ea}{r^3}(m^2-1).$$

Hence the surface density is inversely as the cube of the distance from the inducing point A.

106.] In the case of the two spheres A and B (Fig. 24), whose radii are a and b and potentials P_a and P_b, the distance between their centres being c, we may determine the surface density at any point of the sphere A by considering it as due to the action of a charge aP_a at A, together with that due to the pairs of points B, D and E, F &c., the successive pairs of images.

Putting $r = PB, \quad r_1 = PE, \quad r_2 = PG$, &c.,
we find

$$\sigma = \frac{1}{4\pi} P_a \left[\frac{1}{a} + \frac{b}{r_1^3} \frac{\{(c^2-b^2)^2 - a^2c^2\}}{a^2c} + \&c. \right]$$

$$- \frac{1}{4\pi} P_b \left[\frac{b}{ar^3}(c^2-a^2) + \frac{b^2c^2}{r_2^3(c^2-a^2)} \left\{ \left(\frac{c^2-a^2-b^2}{c^2-a^2} \right)^2 - \frac{a^2}{c^2} \right\} + \&c. \right]$$

If we call B the inducing body and A the induced body, then we may consider the electrification induced on A as consisting of two parts, one depending on the potential of B and the other on its own potential.

The part depending on P_b is called by some writers on electricity the *induced electrification of the first species.* When A is not insulated it constitutes the whole electrification, and if P_b is positive it is negative over every part of the surface, but greatest in numerical nature at the point nearest to B.

The part depending on P_a is called the *induced electrification of the second species.* It can only exist when A is insulated, and it is everywhere of the same sign as P_a. If A is insulated and without charge, then the induced electrifications of the first and second species must be equal and opposite. The surface-density is negative on the side next to B and positive on the side furthest from B, but though the total quantities of positive and negative electrification are equal, the negative electrification is more concentrated than the positive, so that the neutral line which separates the positive from the negative electrification is not the equator of the sphere, but lies nearer to B.

The condition that there shall be both positive and negative electrification on the sphere is that the value of σ at the points nearest to B and farthest from B shall have opposite signs. If a and b are small compared with c, we may neglect all the terms of the coefficients of P_a and P_b after the first. The values of r lie between $c+a$ and $c-a$. Hence, if P_a is between $P_b \frac{b(c-a)}{(c+a)^2}$ and $P_b \frac{b(c+a)}{(c-a)^2}$, there will be both positive and negative electrification on A, divided by a neutral line, but if P_a is beyond these limits, the electrification of every part of the surface will be of one kind; negative if P_a is below the lower limit, and positive if it is above the higher limit.

CHAPTER VIII.

ON ELECTROSTATIC CAPACITY.

107.] THE capacity of a conductor is measured by the charge of electricity which will raise its potential to the value unity, the potential of all other conductors in the field being kept at zero. The capacity of a conductor depends not only on its own form and size, but on the form and position of the other conductors in the field. The nearer the uninsulated conductors are placed the greater is the capacity of the charged conductor.

An apparatus consisting of two insulated conductors, each presenting a large surface to the other with a small distance between them, is called a *condenser*, because a small electromotive force is able to charge such an apparatus with a large quantity of electricity.

The simplest form of condenser, that to which the name is most commonly applied, consists of two disks placed parallel to each other, the medium between them being air. When one of these disks is connected to the zinc and the other to the copper electrode of a voltaic battery, the disks become charged with negative and positive electricity respectively, and the amount of the charge is the greater the nearer the disks are placed to each other, being approximately inversely as the distance between them. Hence by bringing the disks very close to each other, connecting them with the electrodes of the battery and then disconnecting them from the battery, we have two large charges of opposite kinds insulated on the disks. If we now remove one of the disks from the other we do work against the electric attraction which draws them together, and we may thus increase the energy of the system so much that, though the original electromotive force was only that of a single voltaic cell, either of the disks when separated may be raised to so high a potential that the gold leaves of an electrometer connected with it are deflected.

It was in this way that Volta demonstrated that the electrification due to a voltaic cell is of the same kind as that due to friction, the copper electrode being positive with respect to the zinc electrode. In this condenser the capacity of each disk depends principally on the distance between it and the other disk, but it also depends in a smaller degree on the nature of the electric field at the back of the disk.

There are other forms of condensers, however, in which one of the conductors is almost or altogether surrounded by the other. In this case the capacity of the inner conductor is almost or altogether independent of everything but the outer conductor. This is the case in the Leyden jar, and in a cable with a copper core surrounded by an insulator the outside of which is protected by a sheathing of iron wires.

108.] But in most cases the charge of each conductor depends not only on the difference between its potential and that of the other conductor, but also in part on the difference between its potential and that of some other body, such as the earth, or the walls of the room where the experiment is made. The charges of the two conductors may, therefore, in the simpler cases be written

$$Q = K(P-p)+HP, \quad \ldots\ldots\ldots\ldots\ldots\ldots \quad (1)$$

$$q = K(p-P)+hp, \quad \ldots\ldots\ldots\ldots\ldots\ldots\ldots \quad (2)$$

where P and p are the potentials, that of the walls of the room being zero, Q and q the charges of the two conductors respectively, K is the capacity of the condenser in so far as it depends on the mutual relation of the two conductors, and H and h represent those parts of the capacity of each conductor which depend on their relation to external objects, such as the walls of the room.

If we connect the second conductor with the earth we make p zero while Q remains the same, and we get for the new values of P, Q, and q,

$$P_1 = P - \frac{K}{K+H}p, \qquad Q_1 = (K+H)P_1, \qquad q_1 = -KP_1. \ldots\ldots (3)$$

If we now insulate the second conductor and connect the first with the earth we make P zero, and

$$p_2 = -\frac{K}{K+h}P_1, \qquad Q_2 = -Kp_2, \qquad q_2 = (K+h)p_2. \ \ldots\ldots (4)$$

If we again insulate the first conductor and put the second to earth,

$$P_3 = -\frac{K}{K+H}p_2, \quad Q_3 = (K+H)P_3, \quad q_3 = -KP_3 \text{......} \quad (5)$$

From this it appears that if we connect first the one and then the other conductor with the earth the values of the potentials and charges will be diminished in the ratio of $\frac{K^2}{(K+H)(K+h)}$ to unity.

Comparison of two condensers.

109.] Let us suppose the condensers to be Leyden jars having an inner and an outer coating.

Let the inner coating of the first jar and the outer coating of the second be connected with a source of electricity and brought to the potential P, while the outer coating of the first and the inner coating of the second are connected with the earth.

Then if Q_1 and Q_2 are the charges of the inner coatings of the two jars,

$$Q_1 = (K_1+H_1)P, \quad Q_2 = -K_2 P. \text{..............} \quad (7)$$

Now let the outer coatings of both jars be connected with the earth, and let the inner coatings be connected with each other. Required the common potential of the inner coatings.

Here we have

$$p_1' = p_2' = 0,$$

$$Q_1 + Q_2 = Q_1' + Q_2', \text{......................} \quad (8)$$

$$P_1' = P_2' = P', \text{............................} \quad (9)$$

and we have to find P'.

Equation (8) becomes, in virtue of (9),

$$(K_1+H_1-K_2)P = (K_1+H_1+K_2+H_2)P'.$$

If $K_1+H_1 = K_2$ the discharge is complete.

110.] The following method, by which the existence of a determinate relation between the capacities of four condensers may be verified, has been employed by Sir W. Thomson.* It corresponds in electrostatics to Wheatstone's Bridge in current electricity.

In Fig. 25 the condensers are represented as Leyden jars. Two of these, P and Q, are placed with their external coatings in contact with an insulating stand β; the other two, R and S, have their

* Gibson and Barclay.

external coatings connected to the earth. The inner coatings of P and R are permanently connected; so are those of Q and S. In performing the experiment the internal coatings of P and R are first charged to a potential, A, while those of Q and S are charged

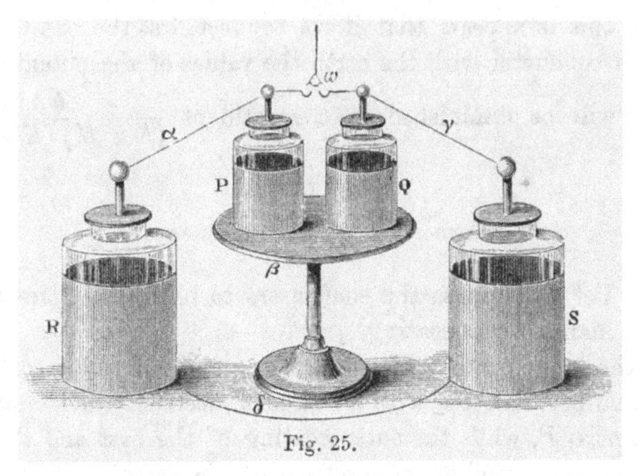

Fig. 25.

to a different potential, C. During this process the stand β is connected to the earth. The stand β is then disconnected from the earth and connected to one electrode of an electrometer, the other

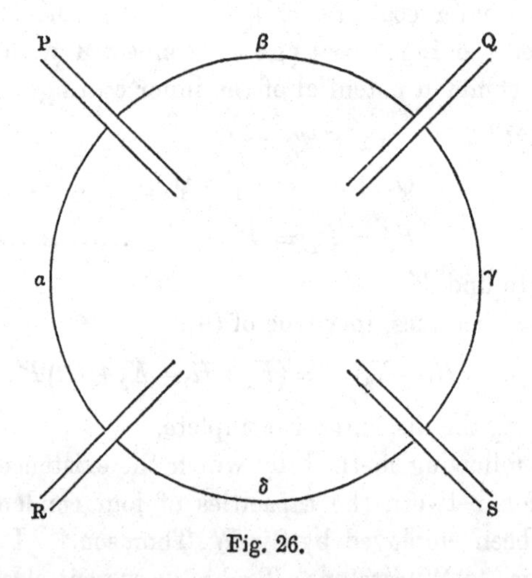

Fig. 26.

electrode being connected to earth. Since β is already reduced to potential zero by connection with the earth, there will be no disturbance of the electrometer unless there is leakage in some of the jars. We shall assume, however, that there is no leakage, and that the electrometer remains at zero.

The inner coatings of the four jars are now made to communicate with each other by dropping the small insulated wire w so as to fall on the two hooks connected with α and γ. Since the potentials of α and γ are different a discharge will occur, and the potential of β will in general be affected, and this will be indicated by the electrometer. If; however, there is a certain relation among the capacities of the jars the potential of β will remain zero.

111.] Let us ascertain what this relation must be. In Fig. 26 the same electrical arrangement is represented under a simpler form, in which the condensers consist each of a pair of disks. Under this form the analogy with Wheatstone's Bridge becomes apparent to the eye. We have to consider the potentials and charges of four conductors. The first consists of the inner coatings of P and R, together with the connecting wire. We shall call this conductor α, its charge a, and its potential A. The second consists of the outer coatings of P and Q, together with the insulating stand β. We shall call this conductor β, its charge b, and its potential B. The third consists of the inner coatings of Q and S and the connecting wire γ. We shall call this γ, its charge c, and its potential C. The fourth consists of the outer coatings of R and S and of the earth with which they are kept connected. We might use the letters δ, d, and D with reference to this conductor, but as its potential is always zero and its charge equal and opposite to that of the other conductors we shall not require to consider it.

The charge of any one of the conductors depends on its own potential together with the potentials of the two adjacent conductors, and also, but in a very slight degree, on that of the opposite conductor.

Let the coefficients of induction between the different pairs of the four conductors be as in the following scheme,—

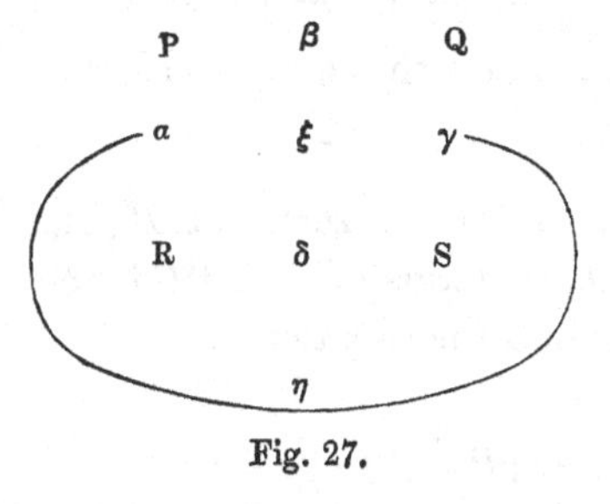

Fig. 27.

in which ξ and η are very small compared with P, Q, R, and S. The coefficient of capacity of any one of the conductors will exceed the sum of its three coefficients of induction by a quantity which will be small if the capacity of the knobs of the jars and their connecting wires are small compared with the whole capacities of the jars. Let us denote this excess by the symbols a, β, γ, δ, which belong to the conductors. The capacities therefore will be,

$$P+R+a+\eta,$$
$$P+Q+\beta+\xi,$$
$$Q+S+\gamma+\eta,$$
$$R+S+\delta+\xi,$$

and the charges will therefore be,

for a, $\quad a=(P+R+a+\eta)A-PB-RD-\eta C,$

for β, $\quad b=(P+Q+\beta+\xi)B-PA-QC-\xi D,$

for γ, $\quad c=(Q+S+\gamma+\eta)C-QB-SD-\eta A,$

for δ, $\quad d=(R+S+\delta+\xi)D-RA-SC-\xi B.$

In the first part of the experiment the potentials of a and γ are A and C respectively, while those of β and δ are zero. Hence, at first,

$$a=(P+R+a+\eta)A-\eta C,$$
$$b=\qquad -PA-QC,$$
$$c=(Q+S+\gamma+\eta)C-\eta A.$$

We need not determine the charge of δ.

Now let a communication be made between a and γ, and let us denote the charges and potentials of the conductors after the discharge by accented letters. The potentials of a and γ will become equal; let us call their common potential y, then

$$A'=C'=y.$$

The sum of their charges remains the same, or

$$a'+c'=a+c.$$

The charge of β remains the same as before, or

$$b'=b,$$

but its potential is no longer zero, but B', and we have to determine the value of B' in terms of A and C by eliminating the other quantities entering into the equations.

After discharge,

$$a'=(P+R+a)y-PB',$$
$$b'=(P+Q+\beta+\xi)B'-(P+Q)y,$$
$$c'=(Q+S+\gamma)y-QB'.$$

Hence, the equation $a' + c' = a + c$ becomes

$$(P+R+Q+S+\alpha+\gamma)y-(P+Q)B' = (P+R+\alpha)A+(Q+S+\gamma)C,$$

and $b' = b$ becomes

$$(P+Q+\beta+\xi)B'-(P+Q)y = -PA-QC.$$

Eliminating y from these equations, we find

$$B'\{(P+Q)(R+S)+(P+Q)(\alpha+\beta+\gamma+\xi)+(R+S+\alpha+\gamma)(\beta+\xi)\}$$
$$= \{Q(R+\alpha)-P(S+\gamma)\}(A-C).$$

If, therefore, the electrometer is not disturbed by the discharge, $B' = O$, and

$$P : Q :: R+\alpha : S+\gamma.$$

CHAPTER IX.

THE ELECTRIC CURRENT.

112.] Let A and B be two metal bodies connected respectively with the inner and outer coatings of a Leyden jar, the inner coating of the jar being positive, so that the potential of A is higher than that of B.

Fig. 28.

Let C be a gilt pith ball suspended by a silk thread. If C is brought into contact with A and B alternately, it will receive a small charge of positive electricity from A every time it touches it, and will communicate positive electricity to B when it touches B.

There will thus be a transference of positive electricity from A to B along the path travelled over by the pith ball, and this is what occurs in every electric current, namely, the passage of electricity along a definite direction. During the motion of the pith ball from A to B it is charged positively, and the electric force between A and B tends to move it in the direction from A to B. After touching B, it becomes charged negatively, so that the electric force, during its return journey, acts from B to A. Hence the ball is acted on by the electric force always in the direction in which it is moving at the time, so that if it is properly suspended the electric force will not only keep up the backward and forward motion, but will communicate to the moving ball an amount of energy which it will expend in a series of rattling blows against the balls A and B. The current of positive electricity from A to B is thus kept up by means of the electromotive force from A to B.

113.] The phenomenon we have been describing may be called a current of Convection. The motion of the electrification takes place in virtue of the motion of the electrified body which *conveys* or *carries* the electricity as it moves from one place to another. But if, instead of the pith ball, we take a metal wire carried by an insulating handle, and cause the two ends of the wire to touch A and B respectively, there will be a transference of electricity from A to B along the wire, though the wire itself does not move.

What takes place in the wire is called a current of Conduction. The effects of the current of conduction on the electrical state of A and B are of precisely the same kind as those of the current of convection. In both cases there is a transference of electrification from one place to another along a continuous path.

In the case of the convection of the charge on the pith ball we may observe the actual motion of the ball, and therefore in this case we may distinguish between the act of carrying a positive charge from A to B and that of carrying a negative charge from B to A, though the electrical effects of these two operations are identical. We may also distinguish between the act of carrying a number of small charges from A to B in rapid succession and with great velocity, and the act of carrying a single great charge, equivalent to the sum of these charges, slowly from A to B in the time occupied by the whole series of journeys in the other case.

But in the case of the current of conduction through a wire we have no reason to suppose that the mode of transference of the charge resembles one of these methods rather than another. All that we know is that a charge of so much electricity is conveyed from A to B in a certain time, but whether this is effected by carrying positive electricity from A to B, or by carrying negative electricity from B to A, or by a combination of both processes, is a question which we have no means of determining. We are equally unable to determine whether the 'velocity of electricity' in the wire is great or small. If there be a substance pervading bodies, the motion of which constitutes an electric current, then the excess of this substance in connexion with a body above a certain normal value constitutes the observed charge of that body, and is a quantity capable of measurement. But we have no means of estimating the normal charge itself. The only evidence we possess is deduced from experiments on the quantity of electricity evolved during the decomposition of one grain of an electrolyte, and this quantity is enormous when compared with any positive or negative charge which we can accumulate within the space occupied by the electrolyte. If, then, the normal charge of a portion of the wire the millionth of an inch in length is equal to the total charge transferred from *A* to *B*, the transference may be effected by the displacement of the electricity in the wire whose linear extent is only the millionth of an inch.

It is therefore quite possible that the velocity of electricity in a telegraph wire may be exceedingly small, less, say, than the hundredth of an inch in an hour, though signals, that is to say, changes in the state of the current, may be propagated along the wire many thousands of miles in a second.

Since, therefore, we are ignorant of the true linear velocity of an electric current, we must measure the *strength* of the current by the quantity of electricity discharged through any section of the conductor in the unit of time, just as engineers measure the discharge of water and gas through pipes, not by the velocity of the water or gas, but by the quantity which passes in a minute.

114.] In many cases we have to consider the whole quantity of electricity which passes rather than the rate at which it passes. This is especially the case when the current lasts only a very short time, or when the current is considered merely as the transition from one permanent state of the system to another. In these cases it is convenient to speak of the total current as the Electric Displacement, the word displacement indicating the final result of a motion without reference to the rate at which it takes place. The passage of a given quantity of electricity along a given path is called an Electric Discharge.

Classification of bodies according to their relation to the transference of electricity.

115.] For the sake of distinction we shall consider a portion of matter whose ends are formed by two equipotential surfaces having different potentials, and whose sides are formed by lines of electric current or displacement.

The ends of the body are called its Electrodes, that at which electricity enters is called the Anode, and that at which it leaves the body is called the Cathode.

The excess of the potential of the anode over that of the cathode is called the External Electromotive Force.

The form of the body may vary from that of a long wire surrounded by air or other insulating matter to that of a thin sheet of the substance, the electricity passing through the thickness of the sheet.

Bodies may be divided into three great classes according to the mode in which they are acted on by electromotive force,—Metals, Electrolytes, and Dielectrics.

First Class.—Metals, &c.

116.] The first class includes all the metals, whether in the solid or liquid state, together with some other substances not regarded by chemists as metals. In these the smallest external electromotive force is capable of producing an electric current, and this current continues to flow as long as the electromotive force continues to act, without producing any change in the chemical properties of the substance. The strength of the permanent current is proportional to the electromotive force. The ratio of the numerical value of the electromotive force to that of the current is called the Resistance of the conductor. The same thing may be otherwise stated by saying that the flow of the current is opposed by an internal electromotive force, proportional to the strength of the current, and to a quantity called the Resistance of the conductor, depending on its form and nature. When the strength of the current is such that this internal electromotive force balances the external electromotive force the current neither increases nor diminishes in strength. It is then said to be a *steady* current.

These relations were first established by Dr. G. S. Ohm, in a work published in 1827. They are expressed by the formula,

$$\text{Electromotive force} = \text{Current} \times \text{Resistance},$$

which is called Ohm's Law.

Generation of Heat by the current.

117.] During the flow of a steady current through a conductor of uniform material of the first class heat is generated in the conductor, but the substance of the conductor will not be affected in any way, for if the heat is allowed to escape as fast as it is generated, the temperature and every other physical condition of the conductor remains the same.

The whole work done by the external electromotive force in urging electricity through the body is therefore spent in generating heat. The dynamical equivalent of the heat generated is therefore equal to the electrical work spent, that is, to the product of the electromotive force into the quantity of electricity transmitted by the current.

Now, the electromotive force is, by Ohm's law, the product of the strength of the current into the resistance, and the quantity of electricity is, by the definition of a current, the product of the current into the time during which it flows, so that we find,

Heat generated measured in dynamical units
$$= \text{Square of Current} \times \text{Resistance} \times \text{Time}.$$

This relation was first established by Dr. Joule, and is therefore called Joule's law. It was also established independently by Lenz.

SECOND CLASS.—ELECTROLYTES.

118.] The second class of substances consists of compound bodies, generally in the liquid form, called Electrolytes.

When an electric current passes through fused chloride of silver, which is an electrolyte, chlorine appears at the anode where the current enters, and silver at the cathode where the current leaves the electrolyte. The quantities of these two substances are such that if combined they would form chloride of silver. The composition of those portions of the electrolyte which lie between the electrodes remains unaltered. Hence, if we fix our attention upon a portion of the electrolyte between two fixed planes perpendicular to the direction of the current, the quantity of silver or of chlorine which enters the portion through one plane must be equal to the quantity which leaves it through the other plane. It follows from this that in every part of the electrolyte the silver is moving in the direction of the current, and the chlorine in the opposite direction.

This operation, in which a compound body is decomposed by an electric current, is called Electrolysis, and the mode in which the current is transmitted is called Electrolytic Conduction. The compound body is called an Electrolyte, and the components into which it is separated are called Ions. That which appears at the anode is called the Anion, and that which appears at the cathode is called the Cation.

The quantity of the substance which is decomposed is proportional to the total quantity of electricity which passes through it, and is independent of the time during which the electricity is passing. The quantity corresponding to the passage of one unit of electricity is called the Electrochemical Equivalent of the substance. Thus, when one unit of electricity is passed through fused chloride of silver, one electrochemical equivalent of silver appears at the cathode and one electrochemical equivalent of chlorine at the anode, and one electrochemical equivalent of chloride of silver disappears.

119.] The electrochemical equivalents of the same substance, as deduced from experiments on different electrolytes which contain it, are consistent with each other. Thus the electrochemical equivalent of chlorine is the same, whether we deduce it from experiments on chloride of silver, or from experiments on hydrochloric acid, and that of silver is the same, whether we deduce it from experiments on chloride of silver, or from experiments on nitrate of silver. These laws of electrolysis were established by Faraday.* If they are strictly true the conduction of electricity through an electrolyte is always electrolytic conduction, that is to say, the electric current is always associated with a flow of the components of the electrolyte in opposite directions.

Such a flow of the components necessarily involves their appearance in a separate form at the anode and the cathode. To effect this separation a certain electromotive force is required depending on the energy of combination of the electrolyte. Thus the electromotive force of one of Daniell's cells is not sufficient to decompose dilute sulphuric acid.

If, therefore, an electrolytic cell, consisting of a vessel of acidulated water, in which two platinum plates are placed as electrodes, is inserted in the circuit of a single Daniell's cell, along with a galvanometer to measure the current, it will be found that though there is a transient current at the instant the circuit is closed, this current rapidly diminishes in intensity, so as to become in a very short time too weak to be measured except by a very sensitive galvanometer.

Neither oxygen nor hydrogen, the chemical components of water, appear in a gaseous form at the electrodes, but the electrodes themselves acquire new properties, showing that a chemical action has taken place at the surface of the platinum plates.

120.] If the Daniell's cell is taken out of the circuit, and the circuit again closed, the galvanometer indicates a current passing through the electrolytic cell in the opposite direction to the original

* *Exp. Res.*, series vii and viii.

current. This current rapidly diminishes in strength and soon vanishes, so that the whole quantity of electricity which is transmitted by it is never greater than that of the primitive current. This reverse current indicates that the platinum plates have acquired a difference of properties by being used as electrodes. They are said to be polarized. The cathode is polarized positively and the anode negatively, so that an electromotive force is exerted in the circuit opposite to that of the Daniell's cell. This electromotive force, which is called the electromotive force of polarization, is the cause of the rapid diminution in the strength of the original current, and of its final cessation.

A chemical examination of the platinum plates shows that a certain quantity of hydrogen has been deposited on the cathode. This hydrogen is not in the ordinary gaseous form, but adheres to the surface of the platinum so firmly that it is not easy to remove the last traces of it.

121.] Faraday's law that conduction takes place in electrolytes only by electrolysis was long supposed not to be strictly true. In the experiment in which a single Daniell's cell furnishes the electromotive force in a circuit containing an electrolyte and a galvanometer, it is found that the current soon becomes very feeble but never entirely vanishes, so that if the electromotive force is maintained long enough, a very considerable quantity of electricity may be passed through the electrolyte without any visible decomposition.

Hence it was argued that electrolytes conduct electricity in two different ways, by electrolysis in a very conspicuous manner and also, but in a very slight degree, in the manner of metals, without decomposition. But Helmholtz has recently* shown that the feeble permanent current can be explained in a different manner, and that we have no evidence that an electrolyte can conduct electricity without electrolysis.

122.] In the case of platinum plates immersed in dilute sulphuric acid, if the liquid is carefully freed from all trace of oxygen or of hydrogen in solution, and if the surfaces of the platinum plates are also freed from adhering oxygen or hydrogen, the current continues only till the platinum plates have become polarized and no permanent current can be detected, even by means of a sensitive galvanometer. When the experiment is made without these pre-

* *Ueber galvanische Polarisation in gasfreien Flüssigkeiten. Monatsbericht d. K. Akad. d. Berlin*, July 1873, p. 587.

cautions, there is generally a certain amount of oxygen or of hydrogen in solution in the liquid, and this, when it comes in contact with the hydrogen or the oxygen adhering to the platinum surface, combines slowly with it, as even the free gases do in presence of platinum. The polarization is thus diminished, and the electromotive force is consequently enabled to keep up a permanent current, by what Helmholtz has called electrolytic convection. Besides this, it is probable that the molecular motion of the liquid may be able occasionally to dislodge molecules of oxygen or of hydrogen adhering to the platinum plates. These molecules when thus absorbed into the liquid will travel according to the ordinary laws of diffusion, for it is only when in chemical combination that their motions are governed by the electromotive force. They will therefore tend to diffuse themselves uniformly through the liquid, and will thus in time reach the opposite electrode, where, in contact with a platinum surface, they combine with and neutralize part of the other constituent adhering to that surface. In this way a constant circulation is kept up, each of the constituents travelling in one direction by electrolysis, and back again by diffusion, so that a permanent current may exist without any visible accumulation of the products of decomposition. We may therefore conclude that the supposed inaccuracy of Faraday's law has not yet been confirmed by experiment.

123.] The verification of Ohm's law as applied to electrolytic conduction is attended with considerable difficulty, because the varying polarization of the electrodes introduces a variable electromotive force, and renders it difficult to ascertain the true electromotive force at any instant. By using electrodes in the form of plates, having an area large compared with the section of the electrolyte, and employing currents alternately in opposite directions, the effect of polarization may be diminished relatively to that of true resistance. It appears from experiments conducted in this way that Ohm's law is true for electrolytes as well as for metals, that is to say, that the current is always proportional to the electromotive force, whatever be the amount of that force. The reason that the external resistance of an electrolyte appears greater for small than for large electromotive forces is that the external electromotive force between the metallic electrodes is not the true electromotive force acting on the electrolyte. There is, in general, a force of polarization acting in the opposite direction to the external electromotive force, and it is only the excess of the external force above the force of polarization that really acts on the electrolyte.

It appears, therefore, that the very smallest electromotive force, if it really acts on the electrolyte, is able to produce conduction by electrolysis. How, then, is this to be reconciled with the fact that in order to produce complete decomposition a very considerable electromotive force is required?

124.] Clausius* has pointed out that on the old theory of electrolysis, according to which the electromotive force was supposed to be the sole agent in tearing asunder the components of the molecules of the electrolyte, there ought to be no decomposition and no current as long as the electromotive force is below a certain value, but that as soon as it has reached this value a vigorous decomposition ought to commence, accompanied by a strong current. This, however, is by no means the case, for the current is strictly proportional to the electromotive force for all values of that force.

Clausius explains this in the following way:—

According to the theory of molecular motion of which he has himself been the chief founder, every molecule of the fluid is moving in an exceedingly irregular manner, being driven first one way and then another by the impacts of other molecules which are also in a state of agitation.

This molecular agitation goes on at all times independently of the action of electromotive force. The diffusion of one fluid through another is brought about by this molecular agitation, which increases in velocity as the temperature rises. The agitation being exceedingly irregular, the encounters of the molecules take place with various degrees of violence, and it is probable that even at low temperature some of the encounters are so violent that one or both of the compound molecules are split up into their constituents. Each of these constituent molecules then knocks about among the rest till it meets with another molecule of the opposite kind and unites with it to form a new molecule of the compound. In every compound, therefore, a certain proportion of the molecules at any instant are broken up into their constituent atoms. At high temperatures the proportion becomes so large as to produce the phenomenon of dissociation studied by M. St. Claire Deville.†

* *Pogg. Ann.* CI. 338 (1857).

† [*Lecons sur la Dissociation, professées devant la Société Chimique.* L. Hachette et Cie. 1866.]

125.] Now Clausius supposes that it is on the constituent molecules in their intervals of freedom that the electromotive force acts, deflecting them slightly from the paths they would otherwise have followed, and causing the positive constituents to travel, on the whole, more in the positive than in the negative direction, and the negative constituents more in the negative direction than in the positive. The electromotive force, therefore, does not produce the disruptions and reunions of the molecules, but finding these disruptions and reunions already going on, it influences the motion of the constituents during their intervals of freedom. The amount of this influence is proportional to the electromotive force when the temperature is given. The higher the temperature, however, the greater the molecular agitation, and the more numerous are the free constituents. Hence the conductivity of electrolytes increases as the temperature rises.

This effect of temperature is the opposite of that which takes place in metals, the resistance of which increases as the temperature rises. This difference of the effect of temperature is sometimes used as a test whether a conductor is of the metallic or the electrolytic kind. The best test, however, is the existence of polarization, for even when the quantity of the free ions is too small to be observed or measured, their presence may be indicated by the electromotive force which they excite.

126.] Kohlrausch * finds that if the electromotive force is one volt per centimetre in length of the electrolyte, then if the electrolyte differs but slightly from pure water at 18° C the velocity of hydrogen is about 0·0029 centimetres per second, and that the actual force on a gramme of hydrogen in the solution required to make it move at the rate of one centimetre per second through the solution is equal to the weight of 330,000,000 kilogrammes.

The velocities of the components of unibasic acids and their salts were found by Kohlrausch to be in the following proportion :—

TABLE I.

H	K	NH_4	Na	Li	$\frac{1}{2}$ Ba	$\frac{1}{2}$ Sr	$\frac{1}{2}$ Ca	$\frac{1}{2}$ Mg
273	48	46	30	19	31	28	24	21

I	Br	Cl	F	NO_3	ClO_3	$C_2H_3O_2$
55	53	50	29	47	36	22

* *Göttingen Nachrichten*, 5 Aug., 1874, 17 May, 1876, and 4 April, 1877.

127.] The specific molecular conductivity (l) of an electrolyte is the sum of the velocities of its components *, and the actual conductivity of any weak solution is found by multiplying the number l by the number of grammes of the substance in a litre and dividing by the molecular weight of the substance, that of hydrogen being 1.

128.] We have reason to believe that water is not an electrolyte, and that it is not a conductor of the electric current. It is exceedingly difficult to obtain water free from foreign matter. Kohlrausch †, however, has obtained water so pure that its resistance was enormous compared with ordinary distilled water. When exposed to the air for [4·3 hours its conductivity rose 70 per cent.], and [in 1060 hours it was increased nearly fortyfold. After long exposure to the air the conductivity was more than doubled in 4·5 hours by the action of tobacco smoke.] Water kept in glass vessels very soon dissolves enough of foreign matter to enable it to conduct freely.

Kohlrausch ‡ has determined the resistance of water containing a very small percentage of different electrolytes, and he finds that the results agree very well with the hypothesis that the velocity with which each ion travels through the liquid is proportional to the electromotive force, the velocity corresponding to unit of electromotive force being different for different ions, but the same for the same ion, whatever the other ion may be with which it is combined. The velocities of different ions in centimetres per second, corresponding to an electromotive force of one volt, are given in Table II.

[TABLE II.

H	K	NH_4	Na	Li	Ba	Sr	Ca	Mg
·0029	·00051	·00049	·00032	·00020	·00033	·00030	·00025	·00022

I	Br	Cl	F	N_2O_3	Cl_2O_3	$C_2H_3O_2$
·00058	·00056	·00053	·00031	·00050	·00038	·00023

]

When the water contains a large percentage of foreign matter the velocities of the ions are no longer the same, as it is no longer through water, but through a liquid of quite different physical properties that they have to make their way. It appears from

* [Compare *Cavendish Papers*, pp. 446, 447.]
† [*Poggendorff, Ergänzungsband*, VIII (1876), pp. 7, 9, 11.]
‡ [*Pogg. Ann.* Vol. CLIV (1875), p 215; Vol. CLIX (1876), p. 242; *Phil. Mag.* June 1875.]

Table III * that though for small percentages of sulphuric acid in water the conducting power is proportional to the percentage of acid, yet as the proportion of acid increases the conducting power increases more slowly till a maximum conducting power is reached, after which the addition of acid diminishes the conducting power †.

TABLE III.

Conductivity of Sulphuric Acid at 18°C *referred to that of Mercury at* 0°C *as unity.*

Percentage of H_2SO_4	10^8K	Percentage of H_2SO_4	10^8K	Percentage of H_2SO_4	10^8K
1	429	60	3487	87	944
2·5	1020	65	2722	88	965
5	1952	70	2016	89	986
10	3665	75	1421	90	1005
15	5084	78	1158	91	1022
20	6108	80	1032	92	1030
25	6710	81	985	93	1024
30	6912	82	947	94	1001
35	6776	83	924	95	958
40	6361	84	915	96	885
45	5766	85	916	97	750
50	5055	86	926	99·4	80
55	4280				

129.] The oxygen and hydrogen which are given off at the electrodes in so many experiments on water containing foreign ingredients are, therefore, not the ions of water separated by strict electrolysis, but secondary products of the electrolysis of the matter in solution. Thus, if the cation is a metal which decomposes water, it unites with an equivalent of oxygen and allows the two equivalents of hydrogen to escape in the form of gas. The anion may be a [compound radicle] which cannot exist in a separate state, [but which exists in the nascent condition, and] contains one equivalent [or more] of [some electronegative element which reacts upon water and liberates oxygen.]

THIRD CLASS.—DIELECTRICS.

130.] The third class of bodies has an electric resistance so much greater than that of metals, or even of electrolytes, that they are often called insulators of electricity. All the gases, many liquids which are not electrolytes, such as spirit of turpentine, naptha, &c., and many solid bodies, such as gutta-percha, caoutchouc in its various forms, amber and resins, crystallized electrolytes, glass when cold, &c., are insulators.

* [See also p. 201.]

† [A similar result was found with nitric acid and some viscous saline solutions.]

They are called insulators because they do not allow a current of electricity to pass through them. They are called dielectrics because certain electrical actions can be transmitted through them. According to the theory adopted in this book, when an electromotive force acts on a dielectric it causes the electricity to be displaced within it in the direction of the electromotive force, the amount of the displacement being proportional to the electromotive force, but depending also on the nature of the dielectric, the displacement due to equal electromotive forces being greater in solid and liquid dielectrics than in air or other gases.

When the electromotive force is increasing, the increase of electric displacement is equivalent to an electric current in the same direction as the electromotive force. When the electromotive force is constant there is still displacement, but no current. When the electromotive force is diminishing, the diminution of the electric displacement is equivalent to a current in the opposite direction.

131.] In a dielectric, electric displacement calls into action an internal electromotive force in a direction opposite to that of the displacement, and tending to reduce the displacement to zero. The seat of this internal force is in every part of the dielectric where displacement exists.

To produce electric displacement in a dielectric requires an expenditure of work measured by half the product of the electromotive force into the electric displacement. This work is stored up as energy within the dielectric, and is the source of the energy of an electrified system which renders it capable of doing mechanical work.

The amount of displacement produced by a given electromotive force is different in different dielectrics. The ratio of the displacement in any dielectric to the displacement in a vacuum due to the same electromotive force is called the Specific Inductive Capacity of the dielectric, or more briefly, the Dielectric Constant. This quantity is greater in dense bodies than in a so-called vacuum, and is approximately equal to the square of the index of refraction. Thus Dr. L. Boltzmann* finds for various substances,

	D.	$\sqrt{D}$.	Index of refraction.
Sulphur (cast)	3·84	1·960	2·040
Colophonium	2·55	1·597	1·543
Paraffin	2·32	1·523	1·536
Ebonite (Hartgummi)	3·15	1·775	

* [*Pogg. Ann.* CLI. (1874), p. 482.]

For a sphere cut from a crystal of sulphur Boltzmann finds D by electrical experiments for the three principal axes, and compares them with the results as calculated from the three indices of refraction.

By electrical experiments	$D_1 = 4{\cdot}773$	$D_2 = 3{\cdot}970$	$D_3 = 3{\cdot}811$
By optical measurements	$D_1 = 4{\cdot}596$	$D_2 = 3{\cdot}886$	$D_3 = 3{\cdot}591$

{Sitzungsb. (Vienna), 9 Jan., 1873.}

132.] Schiller (*Pogg. Ann.* CLII. 535) ascertained the time of the electrical vibrations when a condenser is discharged through an electromagnet. He finds in this way the following values of the dielectric coefficients of various substances, and compares them with those found by Siemens by the method of a rapid commutator.

	Schiller.	Siemens.	μ^2.	μ.
Ebonite (Hartgummi)	2·21	2·76		
Pure rubber	2·12	2·34	2·25	1·50
Vulcanized grey, do.	2·69	2·94		
Paraffin, quick cooled, clear	1·68			
„ slow cooled, milk white	1·81	1·92	2·19	1·48
„ another specimen	1·89	2·47	2·34	1·53
Straw coloured glass	2·96	4·12		
„ „	3·66			
White mirror glass	5·83	6·34		

P. Silow {*Pogg. Ann.* CLVI (1875), [p. 395]}* finds for oil of turpentine

$$D = 2{\cdot}21 \qquad \sqrt{D} = 1{\cdot}490 \qquad \mu_\infty = 1{\cdot}456.$$

Faraday did not succeed in detecting any difference in the dielectric constants of different gases. Dr. Boltzmann † however has succeeded by a very ingenious method in determining it for various gases at 0°C, and at one atmosphere pressure.

	D.	$\sqrt{D}$.	μ.
Air	1·000590	1·000295	1·000294
Carbonic Acid	1·000946	1·000473	1·000449
Hydrogen	1·000264	1·000132	1·000138
Carbonic Oxide	1·000690	1·000345	1·000340
Nitrous Oxide	1·000994	1·000497	1·000503
Olefiant Gas	1·001312	1·000656	1·000678
Marsh Gas	1·000944	1·000472	1·000443

Disruptive Discharge.

133.] If the electromotive force acting at any point of a dielectric is gradually increased, a limit is at length reached at which there

* [See also CLVIII. (1876), pp. 306 *et sqq.*]

† [*Pogg. Ann.* CLI. (1875), p. 403.]

is a sudden electrical discharge through the dielectric, generally accompanied with light and sound. The dielectric, if solid, is often pierced, cracked, or broken, and portions of it are often dispersed in the form of vapour. This phenomenon appears to be analogous to the rupture of a solid body when exposed to a continually increasing stress. This analogy is so complete that we may make use of the same terms in describing the behaviour of media under the action of electromotive force as we apply to bodies under the action of stress. Thus electromotive force and electric displacement correspond to ordinary force and ordinary displacement; the electromotive force which produces disruptive discharge corresponds to the breaking stress. Conduction, or the transmission of electricity, corresponds to permanent bending.

Thus if we consider the twisting of a wire on the one hand, and the transmission of electricity through a body on the other, the moment of the couple which twists the wire will correspond to the electromotive force acting on the body, and the angle through which the wire is twisted will correspond to the electric displacement. If the wire, when the force is removed, returns to its former shape and becomes completely untwisted it is said to be elastic. Such a wire corresponds to a dielectric which acts as a perfect insulator with respect to the electromotive force employed. If the twisting couple is increased up to a certain limit the wire gives way and is broken. This corresponds to disruptive discharge, and the ultimate strength of the wire corresponds to the greatest electromotive force which the dielectric can support, which we may call its electric strength.

If before rupture takes place the wire yields so that it will no longer completely untwist itself when the force is removed it is said to be plastic. It corresponds to a dielectric which conducts electricity to a certain extent.

If no such permanent twist can be given to the wire by a force which is not sufficient to break it, the wire is called brittle. In like manner we may speak of those dielectrics such as air, which will not transmit electricity except by the disruptive discharge, as electrically brittle.

134.] Many wires after being kept twisted for some time and then set free immediately untwist themselves, but through a smaller angle than they were twisted. In the course of time, however, they go on untwisting themselves, but very slowly, the process sometimes going on for days or weeks. In like manner many dielectrics such as the glass of a Leyden jar or the gutta percha of a submarine cable, after being subjected for some time to electromotive force and then placed in a closed circuit give an instantaneous discharge which is less than the original charge. After this discharge, however, they are capable of giving residual discharges which become more and more feeble, and if the circuit is kept closed a quantity of electricity will slowly ooze out through the circuit, the current becoming feebler and feebler as the charge is more nearly exhausted.

Mechanical Illustration of the Properties of a Dielectric.

135*.] Five tubes of equal sectional area A, B, C, D and P are arranged in circuit as in the figure. A, B, C and D are vertical and equal, and P is horizontal.

The lower halves of A, B, C, D are filled with mercury, their upper halves and the horizontal tube P are filled with water.

A tube with a stopcock Q connects the lower part of A and B with that of C and D, and a piston P is made to slide in the horizontal tube.

Let us begin by supposing that the level of the mercury in the four tubes is the same, and that it is indicated by A_0, B_0, C_0, D_0, that the piston is at P_0, and that the stopcock Q is shut.

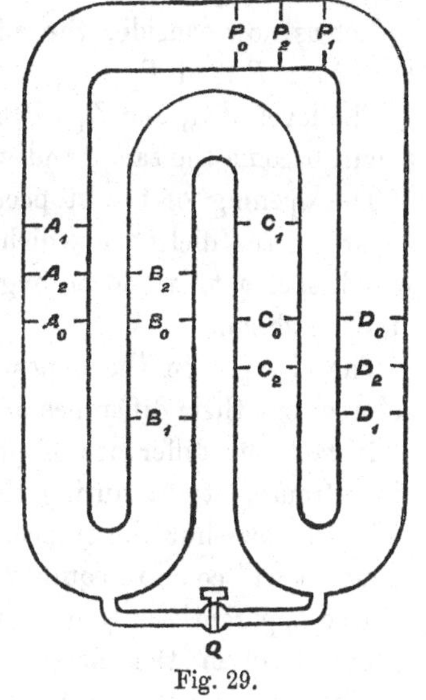

Fig. 29.

Now let the piston be moved from P_0 to P_1, a distance a. Then, since the sections of all the tubes are equal, the level of the mercury in A and C will rise a distance a, or to A_1 and C_1, and the mercury in B and D will sink an equal distance a, or to B_1 and D_1.

The difference of pressure on the two sides of the piston will be represented by $4a$.

This arrangement may serve to represent the state of a dielectric acted on by an electromotive force $4a$.

The excess of water in the tube D may be taken to represent a positive charge of electricity on one side of the dielectric, and the excess of mercury in the tube A may represent the negative charge on the other side. The excess of pressure in the tube P on the side of the piston next D will then represent the excess of potential on the positive side of the dielectric.

If the piston is free to move it will move back to P_0 and be in equilibrium there. This represents the complete discharge of the dielectric.

During the discharge there is reversed motion of the liquids throughout the whole tube, and this represents that change of electric displacement which we have supposed to take place in a dielectric.

I have supposed every part of the system of tubes filled with incompressible liquids, in order to represent the property of all electric displacement that there is no real accumulation of electricity at any place.

Let us now consider the effect of opening the stopcock Q while the piston P is at P_1.

The level of A_1 and D_1 will remain unchanged, but that of B and C will become the same, and will coincide with B_0 and C_0.

The opening of the stopcock Q corresponds to the existence of a part of the dielectric which has a slight conducting power, but which does not extend through the whole dielectric so as to form an open channel.

The charges on the opposite sides of the dielectric remain insulated, but their difference of potential diminishes.

In fact, the difference of pressure on the two sides of the piston sinks from $4a$ to $2a$ during the passage of the fluid through Q.

If we now shut the stopcock Q and allow the piston P to move freely, it will come to equilibrium at a point P_2, and the discharge will be apparently only half of the charge.

The level of the mercury in A and B will be $\frac{1}{2}a$ above its original level, and the level in the tubes C and D will be $\frac{1}{2}a$ below its original level. This is indicated by the levels A_2, B_2, C_2, D_2.

If the piston is now fixed and the stopcock opened, mercury will flow from B to C till the level in the two tubes is again at B_0 and C_0. There will then be a difference of pressure $= a$ on the two sides of the piston P. If the stopcock is then closed and the piston P left free to move, it will again come to equilibrium at a point P_3, half way between P_2 and P_0. This corresponds to the residual charge which is observed when a charged dielectric is first discharged and then left to itself. It gradually recovers part of its charge, and if this is again discharged a third charge is formed, the successive charges diminishing in quantity. In the case of the illustrative experiment each charge is half of the preceding, and the discharges, which are $\frac{1}{2}$, $\frac{1}{4}$, &c. of the original charge, form a series whose sum is equal to the original charge.

If, instead of opening and closing the stopcock, we had allowed it to remain nearly, but not quite, closed during the whole experiment, we should have had a case resembling that of the electrification of a dielectric which is a perfect insulator and yet exhibits the phenomenon called 'electric absorption.'

To represent the case in which there is true conduction through the dielectric we must either make the piston leaky, or we must establish a communication between the top of the tube A and the top of the tube D.

In this way we may construct a mechanical illustration of the properties of a dielectric of any kind, in which the two electricities are represented by two real fluids, and the electric potential is represented by fluid pressure. Charge and discharge are represented by the motion of the piston P, and electromotive force by the resultant force on the piston.

136.] The electric strength of a dielectric medium depends on the nature of the medium and its density and temperature. Thus the electromotive force required to produce a disruptive discharge is greater in glass or ebonite than in air.

The electric strength of air or any other gas may be tested by causing sparks to pass through a portion of the gas between two balls of metal. If the experiment is conducted in a glass vessel from which the air may be exhausted by an air pump, it is found that the electromotive force necessary to produce the discharge diminishes, while the pressure is reduced from that of the atmosphere to that of about 3 millimetres of mercury. If the supply of electricity is kept up at a constant rate, the sparks become smaller and more frequent, till at last there appears to be a continuous flow. If, however, the exhaustion be carried further, the electric strength again increases, till in the most perfect vacuum hitherto made the electromotive force required to produce a spark between electrodes 6 centimetres apart is so great that the discharge does not take place between the electrodes, but passes round the outside of the vessel through a distance of 20 centimetres of air at the ordinary pressure. It would therefore seem as if a perfect vacuum would present an almost insuperable resistance to the passage of electricity. A small quantity of gas, however, introduced into the empty space renders it incapable of withstanding even a small electromotive force. This diminution of the electric strength, however, does not go on when the density of the gas is still further increased, but for pressures of a centimetre and upwards the electric strength increases as the density increases.

137.] The electric strength of air diminishes rapidly as the temperature rises. The heated air which rises from a flame conducts electricity freely. The best way of discharging the electrification of the surface of a solid dielectric is to pass the electrified body over a flame. In most experiments with heated air the air is in motion. It is therefore desirable that experiments should be made on the conductivity of air at various temperatures, contained in a closed vessel and free from currents.

138.] In order to test the insulating properties of air and other gases I made the following experiment :—

A tube half an inch in diameter, *CD*, is supported on an insulated stand *c*. A rod *AB*, a quarter of an inch in diameter, is supported by the insulating stand *a* so that about 6 inches of the rod is within the tube with a cylindrical shell of air about an eighth of an inch thick between it and the inside of the tube. The tube is connected with one electrode of a battery of 50 Leclanché cells, the other electrode being connected to earth. The rod is connected to one electrode of Thomson's quadrant electrometer, the other electrode being connected to earth. A tube, *F*, which is fixed so as not to touch the tube *CD*, is used for sending a current of hot air or steam through the tube *CD*. The part of the tube *CD* which contains the

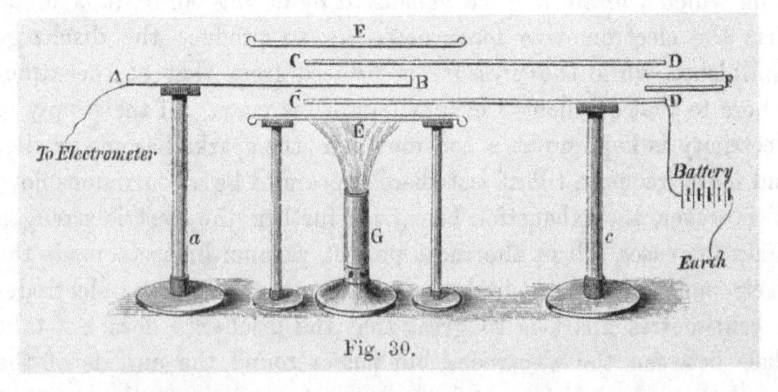

Fig. 30.

rod *AB* is surrounded by a wider tube *E* of thick brass which may be heated by a gas furnace so as to keep the inner tube and rod hot without exposing them to the current of the products of combustion of the burner.

The sensitiveness of this apparatus was shown by the effect of communicating a small charge to the tube *E*. The electrometer was immediately deflected on account of induction between the tube and the rod *AB*. The rod *AB* was then discharged to earth so that the electrometer indicated zero, the tube remaining at a higher potential. If any conduction were now to take place through the air between the tube and the rod it would be indicated by the electrometer. No conduction however could be observed even after the lapse of a quarter of an hour, and when hot air and steam were blown through the tube. At the end of the experiment the tube was discharged to earth, when a negative deflection of the electrometer was observed, shewing that the tube had remained charged during the whole experiment.

139.] Other experiments were afterwards made in which mercury and sodium were made to boil in a bent glass tube while raised to a high potential by a battery of 50 Leclanché cells. A thick copper wire (Fig. 31) was placed on an insulating stand so that the end of the wire was within the glass tube and surrounded by the vapour of the metal. It was necessary that the wire should not be allowed to touch the tube, because glass at a high temperature is a good conductor. It was also necessary to see that the products of combustion from the Bunsen burner did not come in contact with the wire after becoming electrified by the hot tube.

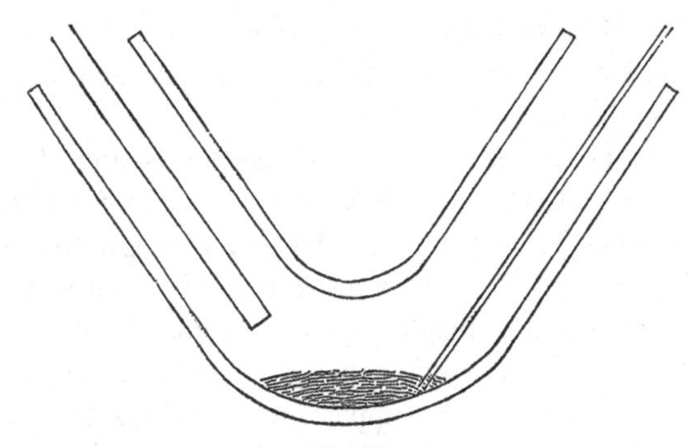

Fig. 31.

The wire was connected with the electrometer, but no evidence of conduction of electricity could be observed, even when the mercury was boiling briskly, and its vapour was being condensed on the wire. But whenever so much mercury had collected on the wire that a drop fell off at the end of the wire, there was a deflection of the electrometer because the drop had become charged by induction from the tube and the removal of this charge affected the electrometer. This however was no evidence of conduction through the metallic vapour, but only indicated that the apparatus was in such a state of electrification that any conduction, if it took place, would produce a sensible indication at the electrometer.

It is difficult to reconcile these experiments on the insulating power of hot gases and vapours with the well-known phenomena of the communication of electricity along the stream of heated matter rising from a flame or even from red-hot metal. This stream acts as a powerful conductor of electricity between the flame and bodies placed at a foot or a yard above it where the temperature of the ascending current is much lower than it was in the experiment of the tube and rod.

140.] The whole theory of the electric properties of gases is in a very imperfect state. According to the kinetic theory of gases, their molecules are in a state of agitation so that they are continually striking against each other. The velocity of this agitation is greater the higher the temperature. It would appear, therefore, that the electric conduction of gases is of the nature of convection. At every collision the whole charge of two of the molecules would be equally divided between them, and thus the tendency of the agitation would be to equalize the charges of all the molecules.

But we can hardly admit a theory of this kind when we consider that we have hitherto obtained no evidence of the conduction of electricity through air at the ordinary pressure and temperature under a feeble electromotive force.

Whenever a body free from projecting points and sharp edges and charged to a low potential is found to lose its charge, the result can always be traced to conduction through the substance or along the surface of the apparatus which is required to support it. The more perfectly insulating we make this apparatus the more slowly does the electrified body lose its charge, so that it is probable that if we could support an electrified body on a perfectly insulating stand so that it could lose its charge only by conduction through the air, it would never lose its charge.

Electric Phenomena of Tourmaline.

141.] Certain crystals of tourmaline and of other minerals possess what may be called Electric Polarity. Suppose a crystal of tourmaline to be at a uniform temperature, and apparently free from electrification on its surface. Let its temperature be now raised, the crystal remaining insulated. One end will be found positively and the other end negatively electrified. Let the surface be deprived of this apparent electrification by means of a flame or otherwise, then if the crystal be made still hotter, electrification of the same kind as before will appear, but if the crystal be cooled the end which was positive when the crystal was heated will become negative.

These electrifications are observed at the extremities of the crystallographic axis. Some crystals are terminated by a six-sided pyramid at one end and by a three-sided pyramid at the other. In these the end having the six-sided pyramid becomes positive when the crystal is heated.

Sir W. Thomson supposes every portion of these and other hemihedral crystals to have a definite electric polarity, the intensity of which depends on the temperature. When the surface is passed through a flame, every part of the surface becomes electrified to such an extent as to exactly neutralize, for all external points, the effect of the internal polarity. The crystal then has no external electrical action, nor any tendency to change its mode of electrification. But if it be heated or cooled the interior polarization of each particle of the crystal is altered, and can no longer be balanced by the superficial electrification, so that there is a resultant external action.

In tourmaline and other pyroelectric crystals it is probable that a state of electric polarization exists, which depends upon temperature, and does not require an external electromotive force to produce it. If the interior of a body were in a state of permanent electric polarisation, the outside would gradually become charged in such a manner as to neutralize the action of the internal electrification for all points outside the body. This external superficial charge could not be detected by any of the ordinary tests, and could not be removed by any of the ordinary methods for discharging superficial electrification. The internal polarization of the substance would therefore never be discovered unless by some means, such as change of temperature, the amount of the internal polarization could be increased or diminished. The external electrification would then be no longer capable of neutralizing the external effect of the internal polarization, and an apparent electrification would be observed, as in the case of tourmaline.

The Electric Glow.

142.] It can be proved by the mathematical theory of electricity that if a conductor having on its surface a sharp conical point is placed in a perfectly insulating medium and electrified, the surface-density of the electricity will increase without limit for points nearer and nearer to the conical point, so that at the conical point itself the surface-density, and therefore the electromotive force, would be infinite. But this result depends on the hypothesis that the air or other surrounding dielectric has an invincible insulating power, which is not the case, and therefore as soon as the electromotive force at the conical point reaches a certain limiting value the insulating power of the air gives way, and there is a disruptive discharge of electricity into the air. A small portion of air close to the conical point thus becomes electrified. The electrified system now consists of the metal conductor with its conical point, together with a rounded mass of electrified air, which covers the point and acts as a sort of sheath to it, so that the boundary of the electrified system is no longer pointed.

This electrified system, if it were solid, would retain its charge, for the electromotive force is not great enough at any place to produce disruptive discharge, but since the air is fluid, and since the electromotive force is greatest in the line of prolongation of the conical point, the electrified particles of air move off in that direction. When they are gone other unelectrified particles take their place round the point, and the point being no longer protected by electrified air, there is another discharge, as at first.

Thus there is continually kept up an influx of uncharged air to the point, a luminous discharge of electricity from the point, called the Electric Glow, and a stream of charged air in the direction of the prolongation of the axis of the cone called the Electric Wind. By checking the influx of air behind the point we may weaken the glow and by increasing the current of air by blowing we may make the glow stronger.

143.] The electric wind which blows from the conical point may be made to drive a little windmill, or if the conductor is made of two wires crossed and having their sharpened ends bent backwards, as in Fig. 32, and supported so as to be capable of rotating, the reaction of the electric wind will make the cross rotate in the direction of the arrows.

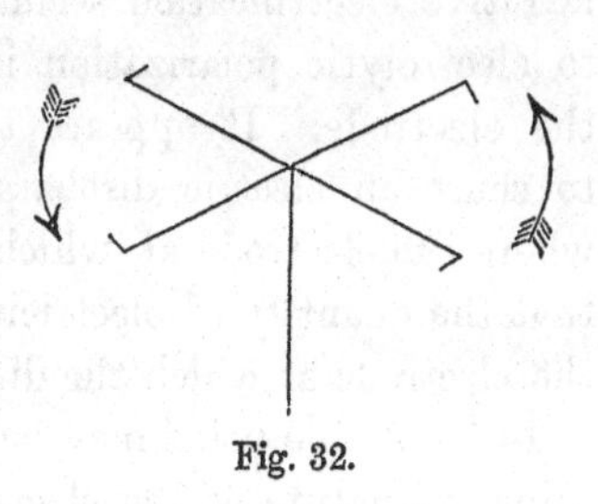
Fig. 32.

It is only close to the electrified point that the motion of the electrified air is in any degree influenced by its electrification. At a short distance from the point the electrified air becomes mixed with other air, and is carried about by the ordinary currents of the atmosphere as an invisible electric cloud.

If we calculate the force due to the electrification of a large body of air at a considerable distance from other electrified bodies, we shall find that it is not capable of producing effects on the motion of so large a mass which are at all comparable to the effects of the slight variations of density and other causes which produce the movements of the atmosphere. Hence the motion of thunder clouds is due almost entirely to atmospheric currents and is not sensibly affected by their electrification.

144.] When an electrified portion of air comes near the surface of a conductor, it induces on that surface an electrification opposite to its own and is attracted towards the surface, but since the electromotive force is small the electrified particles may remain for a long time in the neighbourhood of the conductor without being drawn into contact with it and discharged.

To detect the presence of this electrified atmosphere clinging to a conductor we have only to insulate the conductor and connect it with an electrometer. If we now blow away the electrified air from its surface, the electrometer will indicate the electrification of the conductor itself, which is of course of the opposite kind to that of the electrified air. Thus, if we hold in the hand a hollow metal cylinder over an electrified point, we may electrify the air within it. If we then place it on an insulated stand in connexion with the electrometer the electrometer will remain at zero till the electrified air is removed, which may be done by blowing air through the cylinder. The electrometer will then indicate the electrification of the cylinder, which is of the opposite kind from that of the electrified air.

145.] The glow is more easily formed in rare air than in dense air, and more easily when the point is positive than when it is negative. This and many other differences between positive and negative electrification seem to depend upon a condition analogous to electrolytic polarization in the stratum of air in contact with the electrode. It appears that the electromotive force required to cause an electric discharge to take place is somewhat smaller where the electrode at which the discharge begins is negative, but that the quantity of electricity in each discharge is greatest when the electrode at which the discharge begins is positive.

146.] A fine point may be used instead of a proof plane to determine the nature of the electrification of any part of the surface of a conductor when electricity is induced upon it in presence of another electrified body. For this purpose the point is fixed to the conductor so as to project a few millimetres from its surface. If the part of the surface to which it is fixed is electrified positively the point gives off positive electricity to the air, and the conductor loses positive electricity or gains negative electricity. This may be ascertained either by removing or discharging the inductor and ascertaining the character of the charge of the induced body, or by connecting the induced body with the electrometer and observing the change of potential as the point throws off its electricity.

It has been found that some vegetable thorns, prickles, or spines act more perfectly in throwing off electricity than the finest pointed needles which can be procured.

The action of the point may be assisted by blowing air from a blowpipe over the point, and in this way we may prevent the electrified air from discharging itself on the surface of the inductor.

The Electric Brush.

147.] The electric brush is a phenomenon which may be produced by electrifying a blunt point or a small ball in air so as to produce an electric field in which the tension diminishes as the distance from the ball increases, but not so rapidly as in the case of a sharp point. The brush consists of a succession of discharges, ramifying as they diverge from the ball into the air, and terminating either by charging portions of air or by reaching some other conductor. The brush produces a sound, the pitch of which depends on the interval between the successive discharges, and there is no current of air as in the case of the glow.

The Electric Spark.

148.] When the tension in the space between the two electrodes is considerable all the way between them, as in the case of two balls whose distance is not very great compared with their radii, the discharge, when it occurs, usually takes the form of a spark, by which nearly the whole electrification is discharged at once.

In this case, when any part of the dielectric has given way, the part next to it in the direction of the electric force is put into a state of greater tension, so that it also gives way, and so the discharge proceeds right through the dielectric. We may compare this breaking down of the dielectric to what occurs when we make a little rent perpendicular to the edge of a piece of paper and then apply tension to the paper in the direction of the edge. The paper is torn through, the disruption beginning at the little rent, but diverging occasionally so as to take in weak places in the paper. The electric spark in the same way begins at the point where the electric tension first overcomes the 'electric strength' of the dielectric, and proceeds from that point, in an apparently irregular path, so as to take in other weak points, such as particles of dust floating in the air.

149.] The investigation of the phenomena of the luminous electric discharge has been greatly assisted by the use of the spectroscope. The light of the spark or other discharge is made to fall on the slit of the collimator of the spectroscope, and after being analysed by the prisms is observed through the telescope. The light as thus analysed is found to consist of a great number of bright lines and bands forming what is called the spectrum of the light. By comparing light from different sources it is found that these bright lines may be divided into groups, each group being due to the presence of a particular substance in the medium through which the discharge takes place.

By using the method introduced by Mr. Lockyer of forming an image of the spark upon the slit by means of a lens, we may obtain at one view a comparison of the constituents of the medium which are rendered luminous by the dielectric discharge at the different points of its path. Close to either electrode the lines are principally due to the metal or metals of which that electrode consists. At greater distances these lines become fainter, thinner, and less numerous, but the spectrum belonging to the gas through which the discharge takes place remains visible.

Some of the lines due to the metals appear longer than others, shewing that they can be formed in regions of the spark where others are no longer visible, owing either to a deficiency in the amount of the metallic vapour or to a want of vigour in the electric disturbance. It thus appears that the electric discharge separates an appreciable amount of matter even from the hardest metals and carries the particles through the air to a distance of several millimetres from the surface of the metal. It also appears by a comparison of sparks from different electrodes and through different gases that no part of the light is emitted by any substance common to all the different cases, but that every line is due to one or other of the chemical substances present.

It follows from this that neither the electric fluid, if there be such a substance, nor any etherial medium such as is supposed to pervade all ordinary matter is rendered luminous during the discharge, for if it were so its spectrum would be visible in all discharges.

On Steady Currents.

150*.] In the case of the current between two insulated conductors at different potentials the operation is soon brought to an end by the equalization of the potentials of the two bodies, and the current is therefore essentially a Transient current.

But there are methods by which the difference of potential of the conductors may be maintained constant, in which case the current will continue to flow with uniform strength as a Steady Current.

The Voltaic Battery.

The most convenient method of producing a steady current is by means of the Voltaic Battery.

For the sake of distinctness we shall describe Daniell's Constant Battery:—

A solution of sulphate of zinc is placed in a cell of porous earthenware, and this cell is placed in a vessel containing a saturated solution of sulphate of copper. A piece of zinc is dipped into the sulphate of zinc, and a piece of copper is dipped into the sulphate of copper. Wires are soldered to the zinc and to the copper above the surface of the liquid. This combination is called a cell or element of Daniell's battery. See Art. 193.

151*.] If the cell is insulated by being placed on a non-conducting stand, and if the wire connected with the copper is put in contact with an insulated conductor A, and the wire connected with the zinc is put in contact with B, another insulated conductor of the same metal as A, then it may be shewn by means of a delicate electrometer that the potential of A exceeds that of B by a certain quantity. This difference of potentials is called the Electromotive Force of Daniell's Cell.

If A and B are now disconnected from the cell and put in communication by means of a wire, a transient current passes through the wire from A to B, and the potentials of A and B become equal. A and B may then be charged again by the cell, and the process repeated as long as the cell will work. But if A and B be connected by means of the wire C, and at the same time connected with the battery as before, then the cell will maintain a constant current through C, and also a constant difference of potentials between A and B. This difference will not, as we shall see, be equal to the whole electromotive force of the cell, for part of this force is spent in maintaining the current through the cell itself.

A number of cells placed in series so that the zinc of the first cell is connected by metal with the copper of the second, and so on, is called a Voltaic Battery. The electromotive force of such a battery is the sum of the electromotive forces of the cells of which it is composed. If the battery is insulated it may be charged with electricity as a whole, but the potential of the copper end will always exceed that of the zinc end by the electromotive force of the battery, whatever the absolute value of either of these potentials may be. The cells of the battery may be of very various construction, containing different chemical substances and different metals, provided they are such that chemical action does not go on when no current passes.

152*.] Let us now consider a voltaic battery with its ends insulated from each other. The copper end will be positively or vitreously electrified, and the zinc end will be negatively or resinously electrified.

Let the two ends of the battery be now connected by means of a wire. An electric current will commence, and will in a very short time attain a constant value. It is then said to be a Steady Current.

Magnetic Action of the Current.

153*.] Oersted discovered that a magnet placed near a straight electric current tends to place itself at right angles to the plane passing through the magnet and the current.

If a man were to place his body in the line of the current so that the current from copper through the wire to zinc should flow from his head to his feet, and if he were to direct his face towards the centre of the magnet, then that end of the magnet which tends to point to the north would, when the current flows, tend to point towards the man's right hand. Thus we see that the electric current has a magnetic action which is exerted outside the current, and by which its existence can be ascertained and its intensity measured without breaking the circuit or introducing anything into the current itself.

The amount of the magnetic action has been ascertained to be strictly proportional to the strength of the current as measured by the products of electrolysis in the voltameter, and to be quite independent of the nature of the conductor in which the current is flowing, whether it be a metal or an electrolyte.

154*.] An instrument which indicates the strength of an electric current by its magnetic effects is called a Galvanometer.

Galvanometers in general consist of one or more coils of silk-covered wire within which a magnet is suspended with its axis horizontal. When a current is passed through the wire the magnet tends to set itself with its axis perpendicular to the plane of the coils. If we suppose the plane of the coils to be placed parallel to the plane of the earth's equator, and the current to flow round the coil from east to west in the direction of the apparent motion of the sun, then the magnet within will tend to set itself with its magnetization in the same direction as that of the earth considered as a great magnet, the north pole of the earth being similar to that end of the compass needle which points south.

The galvanometer is the most convenient instrument for measuring the strength of electric currents. We shall therefore assume the possibility of constructing such an instrument in studying the laws of these currents, and when we say that an electric current is of a certain strength we suppose that the measurement is effected by the galvanometer.

On Systems of Linear Conductors.

155*.] Any conductor may be treated as a linear conductor if it is arranged so that the current must always pass in the same manner between two portions of its surface which are called its electrodes. For instance, a mass of metal of any form the surface of which is entirely covered with insulating material except at two places, at which the exposed surface of the conductor is in metallic contact with electrodes formed of a perfectly conducting material, may be treated as a linear conductor. For if the current be made to enter at one of these electrodes and escape at the other the lines of flow will be determinate, and the relation between electromotive force, current and resistance will be expressed by Ohm's Law, for the current in every part of the mass will be a linear function of E. But if there be more possible electrodes than two, the conductor may have more than one independent current through it.

Ohm's Law.

156*.] Let E be the electromotive force in a linear conductor from the electrode A_1 to the electrode A_2. (See Art. 5.) Let C be the strength of the electric current along the conductor, that is to say, let C units of electricity pass across every section in the direction $A_1 A_2$ in unit of time, and let R be the resistance of the conductor, then the expression of Ohm's Law is

$$E = CR. \qquad (1)$$

The Resistance of a conductor is defined to be the ratio of the electromotive force to the strength of the current which it produces. The introduction of this term would have been of no scientific value unless Ohm had shewn, as he did experimentally, that it corresponds to a real physical quantity, that is, that it has a definite value which is altered only when the nature of the conductor is altered.

In the first place, then, the resistance of a conductor is independent of the strength of the current flowing through it.

In the second place the resistance is independent of the electric potential at which the conductor is maintained, and of the density of the distribution of electricity on the surface of the conductor.

It depends entirely on the nature of the material of which the conductor is composed, the state of aggregation of its parts and its temperature.

The resistance of a conductor may be measured to within one ten thousandth or even one hundred thousandth part of its value, and so many conductors have been tested that our assurance of the truth of Ohm's Law is now very high*.

Linear Conductors arranged in Series.

157*.] Let A_1, A_2 be the electrodes of the first conductor and let the second conductor be placed with one of its electrodes in contact with A_2, so that the second conductor has for its electrodes A_2, A_3. The electrodes of the third conductor may be denoted by A_3 and A_4.

Let the electromotive force along each of these conductors be denoted by E_{12}, E_{23}, E_{34}, and so on for the other conductors.

Let the resistance of the conductors be

$$R_{12}, \quad R_{23}, \quad R_{34}, \text{ \&c.}$$

Then, since the conductors are arranged in series so that the same current C flows through each, we have by Ohm's Law,

$$E_{12} = CR_{12}, \quad E_{23} = CR_{23}, \quad E_{34} = CR_{34}. \quad \text{........} \quad (2)$$

If E is the resultant electromotive force, and R the resultant resistance of the system, we must have by Ohm's Law,

$$E = CR. \quad \text{..........................} \quad (3)$$

Now $\quad E = E_{12} + E_{23} + E_{34}, \quad$ .... (4)

the sum of the separate electromotive forces,

$$= C(R_{12} + R_{23} + R_{34}) \quad \text{by equations (2).}$$

Comparing this result with (3), we find

$$R = R_{12} + R_{23} + R_{34}. \quad \text{......................} \quad (5)$$

Or, *the resistance of a series of conductors is the sum of the resistances of the conductors taken separately.*

Potential at any Point of the Series.

Let A and C be the electrodes of the series, B a point between them, a, c, and b the potentials of these points respectively. Let R_1 be the resistance of the part from A to B, R_2 that of the part from B to C, and R that of the whole from A to C, then, since

$$a - b = R_1 C, \quad b - c = R_2 C, \quad \text{and} \quad a - c = RC,$$

the potential at B is

$$b = \frac{R_2 a + R_1 c}{R}, \quad \text{......................} \quad (6)$$

* [See Report of British Association, 1876.]

which determines the potential at B when those at A and C are given.

Resistance of a Multiple Conductor.

158*.] Let a number of conductors ABZ, ACZ, ADZ be arranged side by side with their extremities in contact with the same two points A and Z. They are then said to be arranged in multiple arc.

Let the resistances of these conductors be R_1, R_2, R_3 respectively, and the currents C_1, C_2, C_3, and let the resistance of the multiple conductor be R, and the total current C. Then, since the potentials at A and Z are the same for all the conductors, they have the same difference, which we may call E. We then have

$$E = C_1 R_1 = C_2 R_2 = C_3 R_3 = CR,$$

but

$$C = C_1 + C_2 + C_3,$$

whence

$$\frac{1}{R} = \frac{1}{R_1} + \frac{1}{R_2} + \frac{1}{R_3}. \qquad (7)$$

Or, *the reciprocal of the resistance of a multiple conductor is the sum of the reciprocals of the component conductors.*

If we call the reciprocal of the resistance of a conductor the conductivity of the conductor, then we may say that *the conductivity of a multiple conductor is the sum of the conductivities of the component conductors.*

Current in any Branch of a Multiple Conductor.

From the equations of the preceding article, it appears that if C_1 is the current in any branch of the multiple conductor, and R_1 the resistance of that branch,

$$C_1 = C\frac{R}{R_1}, \qquad (8)$$

where C is the total current, and R is the resistance of the multiple conductor as previously determined.

Kirchhoff has stated the conditions of a linear system in the following manner, in which the consideration of the potential is avoided.

(1) (Condition of 'continuity.') At any point of the system the sum of all the currents which flow towards that point is zero.

(2) In any complete circuit formed by the conductors the sum of the electromotive forces taken round the circuit is equal to the

sum of the products of the currents in each conductor multiplied by the resistance of that conductor.

Longitudinal Resistance of Conductors of Uniform Section.

159*.] Let the resistance of a cube of a given material to a current parallel to one of its edges be ρ, the side of the cube being unit of length, ρ is called the 'specific resistance of that material for unit of volume.'

Consider next a prismatic conductor of the same material whose length is l, and whose section is unity. This is equivalent to l cubes arranged in series. The resistance of the conductor is there fore $l\rho$.

Finally, consider a conductor of length l and uniform section s. This is equivalent to s conductors similar to the last arranged in multiple arc. The resistance of this conductor is therefore

$$R = \frac{l\rho}{s}.$$

When we know the resistance of a uniform wire we can determine the specific resistance of the material of which it is made if we can measure its length and its section.

The sectional area of small wires is most accurately determined by calculation from the length, weight, and specific gravity of the specimen. The determination of the specific gravity is sometimes inconvenient, and in such cases the resistance of a wire of unit length and unit mass is used as the 'specific resistance per unit of weight.'

If r is this resistance, l the length, and m the mass of a wire, then

$$R = \frac{l^2 r}{m}.$$

CHAPTER X.

PHENOMENA OF AN ELECTRIC CURRENT WHICH FLOWS THROUGH HETEROGENEOUS MEDIA.

1. *Thermo-electric phenomena.*

160.] Seebeck, in 1822, discovered that if a circuit is formed of two different metals, and if the two junctions of the metals are kept at different temperatures, an electric current tends to flow round the circuit. If the metals are iron and copper at temperatures below 280° C., the current flows from copper to iron through the hotter junction. There is therefore, in general, an electromotive force acting in a definite direction round the circuit, whenever the two junctions are at different temperatures.

In a circuit formed of any number of metals all at the same temperature, there can be no current, for if there were a current it might be constantly employed to work a machine or to generate heat in a conductor, and this without any energy being supplied to the system from without, for in order to keep the circuit at a constant temperature nothing is required except to prevent heat from entering or leaving it. Hence at any given temperature the electromotive force in a circuit of three metals, A, B, C must be zero for the whole circuit. Hence if the electromotive force from C to A is a, and that from C to B is b, and that from B to A x, then in the circuit A, B, C, the total electromotive force is $a-b-x=0$, so that x, the electromotive force from B to A is represented by $a-b$, where a and b are quantities determined by observation of the electromotive force from any third metal C to the metals A and B. We may express this by saying that the quantities a and b are the *potentials* of the metals A and B with respect to a third metal C at the given temperature. The potential of A with respect to B is $a-b$. The actual determination of the relative potentials of the metals will be explained in Art. 182.

161.] It has been shewn by Magnus* that if a circuit be formed of a single metal, no current will be formed in it, however the temperature and the section of the conducting circuit may vary in different parts. Since in this case there is necessarily conduction of heat, and consequently dissipation of energy, we cannot, as in the former case, consider the result as self-evident. The electromotive force, for instance, between two portions of the circuit at given temperatures might depend on the length or the mode of variation of the section of the intermediate portion of the circuit. In fact the experiments of Le Roux and others have shewn that the law of Magnus is no longer applicable in a circuit in which there is a very abrupt variation of temperature, as at the instant when the circuit is closed by a hot wire coming in contact with a cold wire of the same metal. Even without

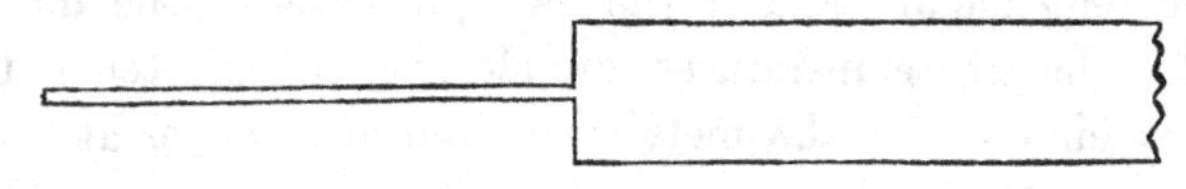

Fig. 33.

any physical discontinuity of the circuit such as is implied in the contact of two separate pieces of wire, a sufficiently abrupt variation of temperature may be produced by taking a thick wire and filing down a certain length of it till it is very thin. If the junction of the thick and the thin portions is placed in a flame, the thin portion will be heated so much more rapidly than the thick portion, that the variation of temperature will be so abrupt that the law of Magnus fails, and we obtain a current in a circuit of one metal; we must therefore modify the statement of the law of Magnus as follows:—

The electromotive force from one point of a conductor of homogeneous metal to another depends only on the temperature of these points unless at any part of the conductor a sensible variation of temperature occurs between points whose distance is within the limits of molecular action.

Thermo-electric power of a metal at a given temperature.

162.] Let us now consider a linear circuit made up of alternate pieces of two metals, say lead and iron. We shall assume lead to be the standard metal, and study the properties of iron in relation to lead.

In the figure the pieces of iron are distinguished by shading. Let the temperatures of the junctions be those indicated in the

* [*Pogg. Ann.* 1851.]

figure, in which the temperatures of the extremities of each piece of iron differ by one degree, but the temperatures of the extremities of each of the intermediate pieces of lead are equal. The total electromotive force round the circuit is the sum of the electromotive

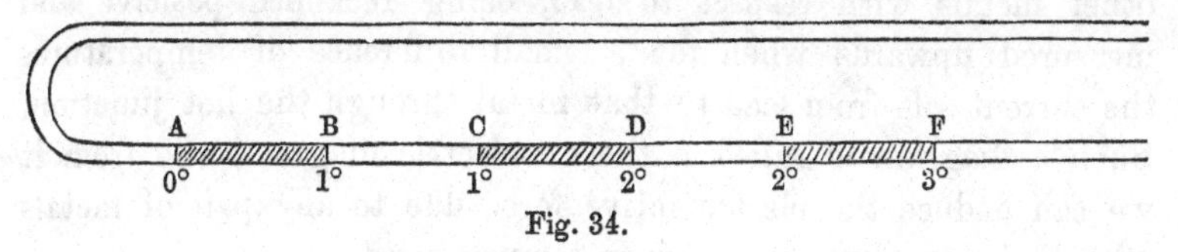

Fig. 34.

forces due to the thermo-electric action of the different pairs of junctions. Now if we consider the pairs A and B, C and D, E and F belonging to the pieces of iron we find that the temperature rises one degree in each piece, but if we take the pairs B and C, D and E belonging to the pieces of lead, the temperature in each piece is uniform and therefore there is no electromotive force in these pieces. We may therefore leave the intermediate pieces of lead out of account, and consider the electromotive force due to the junctions A and F as equivalent to the sum of the electromotive forces of the three pairs of junctions A and B, C and D, E and F.

Hence if a diagram is constructed in which the axis OZ is marked with the degrees of the thermometric scale and in which the area $0°PQ1°$ represents the electromotive force when the junctions are at 0° and 1° and so on, then the electromotive force

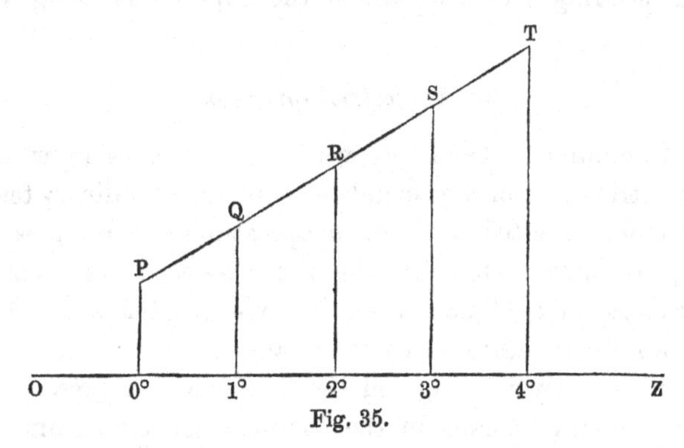

Fig. 35.

when the junctions are at any given temperatures will be represented by the area included between the axis, the ordinates at the given temperatures and the line $PQRST$.

163.] Any ordinate such as $0°P$, $1°Q$, &c., is called the Thermoelectric Power of iron with respect to lead at 0°, 1°, &c., and is reckoned positive when, for a small difference of temperature, the current is from lead to iron through the hot junction.

We may also on the same diagram construct other lines, the ordinates of which represent the thermo-electric powers of any other metals with respect to lead, being reckoned positive and measured upwards when for a small difference of temperatures the current sets from lead to that metal through the hot junction. Such a diagram is called a thermo-electric diagram, and from it we can deduce the electromotive force due to any pair of metals with their junctions at any given temperatures.

Thus if aA is the line representing the metal A, and bB another representing the metal B, and T, t the temperatures of the junctions, the electromotive force of the circuit is represented by the area $ABbaA$ and it acts in the direction indicated, namely, from the metal A to the metal B through the hot junction.

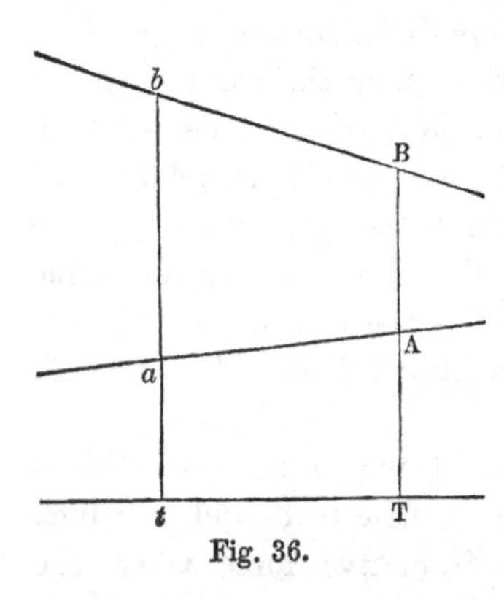

Fig. 36.

If, instead of lead, we had assumed any other metal as the standard metal, the diagram would have been altered in form, but all areas measured on the diagram would have remained the same, the change of form being due to a shearing strain in which the slipping is along vertical lines.

Thermo-electric Inversion.

164.] Cumming in 1823 discovered several cases in which the thermo-electric order of two metals as observed at ordinary temperatures becomes inverted at high temperatures. The lines corresponding to these metals on the thermo-electric diagram must therefore cross one another at some intermediate temperature, called the Neutral Temperature for these metals.

Tait has recently investigated the lines which represent a considerable number of metals in the thermo-electric diagram, and he finds that for most metals they are nearly if not exactly straight lines. The lines for iron and nickel however have considerable sinuosities, so that they may intersect the straight lines belonging to another metal in several different points corresponding to several different neutral temperatures.

Thermal effects of the Current.

165.] By applying the principle of the conservation of energy to the case of a thermo-electric current, it is easy to see that certain thermal effects must accompany the electric current.

Let us consider what takes place while one unit of electricity is transmitted across any section of the circuit. The work done on the electric current is the product of the electromotive force into the quantity of electricity transmitted, and since this latter quantity is unity, the work is numerically equal to the electromotive force, and is represented by the area *ABba* in the thermo-electric diagram. If the current is allowed to flow without anything to impede it except the resistance of the circuit, the whole of the work will be converted into heat, but if the resistance of any part of the circuit such as a long and fine wire greatly exceeds that of the thermo-electric couple, the heat generated in that part of the circuit will greatly exceed that generated in the thermo-electric couple itself. Instead of allowing the current to generate heat, we may make it drive a magneto-electric engine, and so convert any given proportion of the work into mechanical work.

Thus for every unit of electricity which is transmitted, a certain amount of work is done by the thermo-electric forces on the current. The only source of this work is the heat of the thermo-electric couple, and therefore, by the principle of the conservation of energy, we conclude that an amount of heat, dynamically equivalent to this work, must have disappeared in some part of the circuit.

166.] Now Peltier* in 1834 found that when an electric current is made to pass from one metal to another which has a higher thermo-electric power, the junction is cooled, or, since there is no permanent change in the metals, there is a disappearance of heat. When the current is made to flow in the opposite direction the junction is heated, indicating a generation of heat.

This thermal effect of the current at the junction is of quite a different kind from the ordinary generation of heat by the current while it overcomes the resistance of a conductor. The latter, which we may call with Thomson the *frictional* generation of heat, is the same when the direction of the current is reversed, and varies as the square of the strength of the current. The former,

* Annales de Chimie et de Physique, lvi. p. 371 (1834).

which we shall call the Peltier effect, is reversed when the current is reversed, and depends simply on the strength of the current.

167.] But Thomson has shewn that besides the Peltier effect, there must in certain metals be another reversible thermal effect of the current. The current must generate or absorb heat when it passes from hotter to colder or from colder to hotter parts of the same metal. Thus, let a thermo-electric couple of copper and iron be kept with one junction *AB* at the neutral temperature which is about 280° C., and the other, *ab*, at some lower temperature. The thermo-electric current is from copper to iron at the hot junction *AB* and from iron to copper at the cold junction *ab*.

Now the Peltier effect at the hot junction, *AB*, is zero, for that junction is at the neutral temperature, and the Peltier effect at the cold junction, *ab*, is a generation of heat, for the current is there passing from the metal of higher to the metal of lower thermo-electric power. Hence the absorption of heat which must exist in order to account for the work done by the current must take place in some other part of the circuit, either in the copper where the current is flowing from cold to hot, or in the iron where it is flowing from hot to cold, or in both metals. This thermal effect of the current was predicted by Thomson as the result of reasoning similar to that here given. He afterwards verified this prediction experimentally, and found that in iron unequally heated a current from hot to cold cools the metal, while a current from cold to hot heats it, and that the reverse thermal effect takes place in copper. We shall refer to this thermal effect as the Thomson effect.

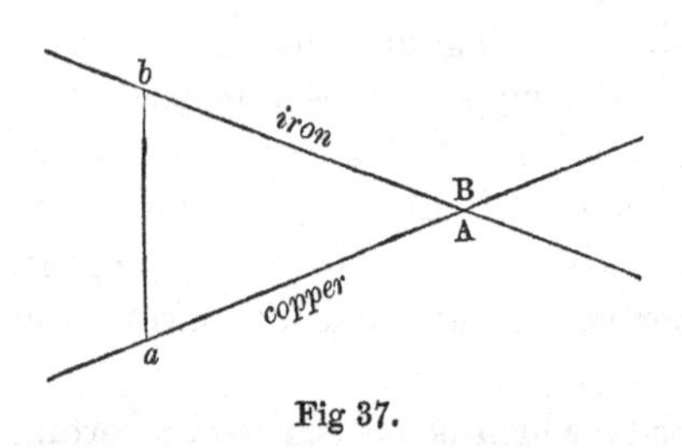

Fig 37.

168.] Thomson has shewn that a very close analogy subsists between these thermo-electric phenomena and those of a fluid circulating in a tube consisting of two vertical branches connected by two horizontal branches. A fluid, heated in one part of the circuit, and passing on into cooler parts of the system, will give out heat, and when it passes from colder to warmer parts will absorb heat, the amount of heat emitted or absorbed depending on the specific heat of the fluid. According to this analogy, positive or vitreous electricity carries heat with it in copper as if it were a real fluid, but in iron it behaves as if its specific heat were a negative quantity which cannot be the case in a real fluid. Hence Thomson expresses the fact by saying that negative or resinous electricity carries heat with it in iron. Neither kind of electricity, therefore, can be regarded in this respect as a real fluid. We may therefore adhere to the usual convention, and speaking of the positive electricity only, we may say that in copper it behaves as if its specific heat were positive, and in iron as if it were negative.

169.] M. Le Roux,* who has made some very careful experiments on the Thomson effect, finds that in lead the specific heat of electricity is either zero or very small indeed. Professor Tait has therefore adopted lead as the standard metal in his thermo-electric measurements.

170.] We may express both the Peltier and the Thomson effects by stating that when an electric current is flowing from places of smaller to places of greater thermo-electric power, heat is absorbed, and when it is flowing in the reverse direction heat is generated, and this, whether the difference of thermo-electric power in the two places arises from a difference in the nature of the metal or from a difference of temperature in the same metal.

171.] The amount of heat absorbed corresponding to a given

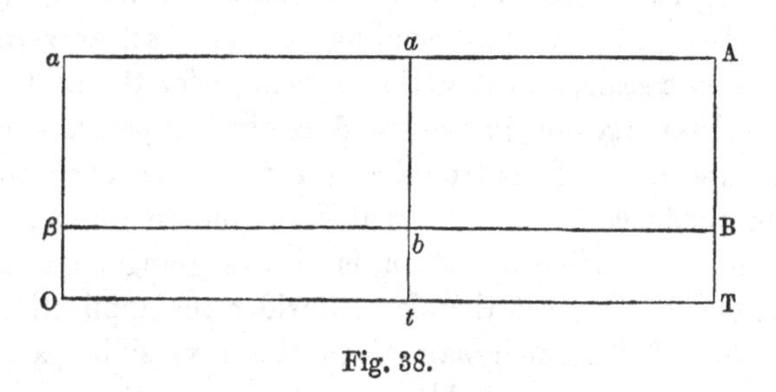

Fig. 38.

increase of thermo-electric power, must depend on the temperature as well as on the amount of that increase. For consider a circuit consisting of two metals, neither of which exhibits the Thomson effect. Such a circuit would be represented in the thermo-electric diagram by the parallelogram *AabB* with horizontal and vertical sides. If the current flows in the direction *AabB* heat is absorbed in *BA* and generated on *ab*, and no reversible thermal effect occurs elsewhere. Also the heat absorbed in *BA* exceeds that generated

* Annales de Chimie et de Physique (4), x. p. 243 (1867)

in ab by a quantity represented by the parallelogram $BAab$. Hence if we produce Aa and Bb and draw the vertical line $\alpha\beta$ at such a distance that the heat absorbed at the junction AB is represented by the parallelogram $BA\alpha\beta$, the heat generated at the junction ab, which, as we have seen, is less than this by the parallelogram $BAab$, will be represented by the parallelogram $ab\beta\alpha$. The Peltier effect therefore is measured by the product of the increase of thermo-electric power in passing from the first metal to the second into the temperature reckoned from some point lower than any observed temperature, and is of the form $(\phi_2-\phi_1)(t-t_1)$, when the current flows from a metal in which the thermo-electric power is ϕ_1 to a metal in which it is ϕ_2, and t is the thermometer reading, and t_1 is a constant, the value of which can be ascertained only by experiment.

172.] Thus far we are led by the principle of the Conservation of Energy. It is a consequence, however, of the Second Law of Thermodynamics, that in all strictly reversible operations in which heat is transformed into work or work into heat, the amount of heat absorbed or emitted at the higher temperature is to that emitted or absorbed at the lower temperature as the higher temperature is to the lower temperature, both being reckoned from absolute zero of the thermodynamic scale. It follows that the line $\alpha\beta$ must be drawn in the position corresponding to the absolute zero of the thermodynamic scale, and that the expression for the heat absorbed may be written $(\phi_2-\phi_1)\theta$, where θ is the temperature reckoned from absolute zero. It is true that the thermo-electric operations cannot be made completely reversible, as the conduction of heat, which is an irreversible operation, is always going on, and cannot be prevented. We must therefore consider the application of the Second Law of Thermodynamics to the reversible part of the phenomena as a very probable conjecture consistent with other parts of the theory of heat, and verified approximately by the measurements of the Peltier and Thomson effects by Le Roux.

173.] We are now able to express all the thermal and electromotive effects in terms of the areas in the thermo-electric diagram. Let Ii be the line for one metal, say iron, Cc that for another, say copper. Let T be the higher temperature and t the lower, and let O represent the position of absolute zero. Let the current flow in the direction $CIic$ till one unit of electricity has passed. Then the heat absorbed at the hot junction will be represented by the area $CIQR$. This is the Peltier effect.

The heat absorbed in the iron is represented by $IiPQ$...Thomson effect.

The heat generated in the cold junction, by $icSP$...Peltier effect.

The heat absorbed in the copper, by $cCRS$...Thomson effect.

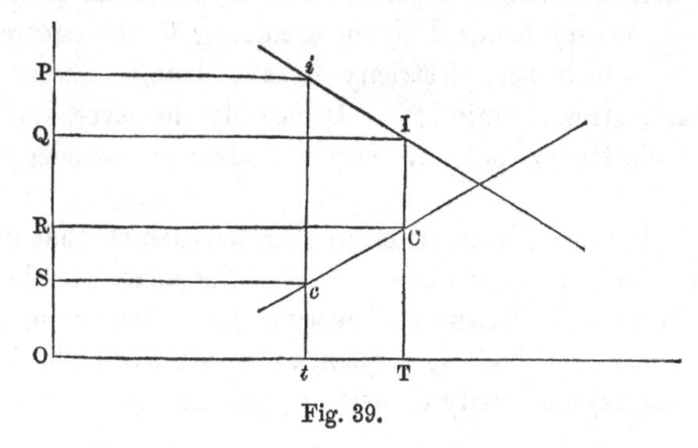

Fig. 39.

The whole heat absorbed is therefore represented by $CIiPSc$, and the heat generated by $icSP$, leaving $CIic$ for the heat absorbed as the result of the whole operation. This heat is converted into the work done on the electric current.

174.] Entropy*, in Thermodynamics, is a quantity relating to a body such that its increase or diminution implies that heat has entered or left the body. The amount of heat which enters or leaves the body is measured by the product of the increase or diminution of entropy into the temperature at which it takes place.

In this treatise we have avoided making any assumption that electricity is a body or that it is not a body, and we must also avoid any statement which might suggest that, like a body, electricity may receive or emit heat.

We may, however, without any such assumption, make use of the idea of entropy, introduced by Clausius and Rankine into the theory of heat, and extend it to certain thermo-electric phenomena, always remembering that entropy is not a thing but a mere instrument of scientific thought, by which we are enabled to express in a

* [Arts. 174–181 consist principally of a repetition of Arts. 167–173, but expressed in the language of the doctrine of Entropy. It was probably the intention of Professor Clerk Maxwell to insert them or some modification of them in place of the foregoing Articles, but it has been thought best not to alter the continuous MS., but simply to insert the separate Articles here as representing a slightly different method of applying the Second Law of Thermodynamics to thermo-electric phenomena.]

compact and convenient manner the conditions under which heat is emitted or absorbed.

175.] When an electric current passes from one metal to another heat is emitted or absorbed at the junction of the metals. We shall therefore suppose that the electric entropy has diminished or increased when the electricity passes from the one metal to the other, the electric entropy being different according to the nature of the medium in which the electricity is, and being affected by its temperature, stress, strain, &c. It is only, however, during the motion of electricity that any thermo-electric phenomena are produced.

176.] It is proved in treatises on thermodynamics that in all reversible thermal operations, what is called the entropy of the system remains the same. (Maxwell's Theory of Heat, 5th ed. p. 190.)

The entropy of a body is a quantity which when the body receives (or emits) a quantity of heat, H, increases (or diminishes) by a quantity $\frac{H}{\theta}$, where θ is the temperature reckoned on the thermodynamic scale. The entropy of a material system is the sum of the entropies of its parts.

177.] The thermal effects of electric currents are in part reversible and in part irreversible, but the reversible effects, such as those discovered by Peltier and Thomson, are always small compared with the irreversible effects—the frictional generation of heat and the diffusion of heat by conduction. Hence we cannot extend the demonstration of the theorem, which applies to completely reversible thermal operations, to thermo-electric phenomena.

But, as Sir Wm. Thomson has pointed out, we have great reason to conjecture that the reversible portion of the thermo-electric effects are subject to the same condition as other reversible thermal operations. This conjecture has not hitherto been disproved by any experiments, and it may hereafter be verified by careful electric and calorimetric measurements. In the meantime the consequences which flow from this conjecture may be conveniently described by an extension of the term entropy to electric phenomena.

The term Electric Entropy, as we shall use it, corresponds to the term Thermo electric Power, as defined by Sir W. Thomson in his fifth paper on the Dynamical Theory of Heat (Trans. R. S. E. 1st May, 1854; Art. 140, p. 151).

Thermo-electric Diagram.

178.] The most convenient method of studying the theory of thermo-electric phenomena is by means of a diagram in which the temperature and electric entropy of a metal at any instant are represented by the horizontal and vertical coordinates of a point on the diagram. Thus, if OC represents the temperature, reckoned from absolute zero on the thermodynamic scale, of a piece of a certain metal, and if CA represents the electric entropy corresponding to the same piece of metal, then the point A will indicate by its position in the diagram the thermo-electric state of the piece of metal. In the same way we may find other points in the diagram corresponding to the same metal under other conditions or to other metals.

If in the path of an electric current electricity passes from one metal to another or from one portion of a metal to another at a different temperature, the different points of the electric circuit

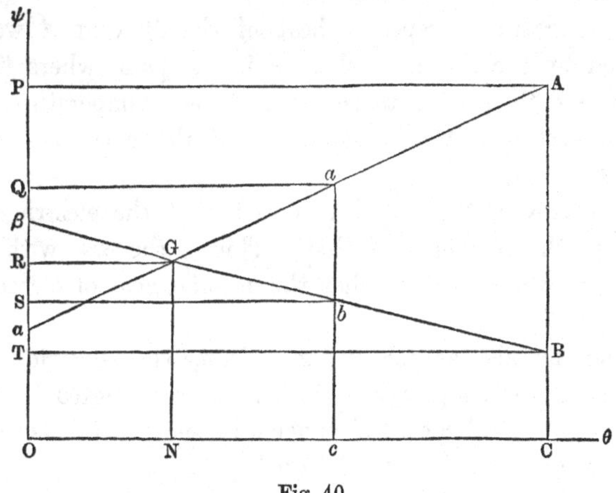

Fig. 40.

will be represented by corresponding points on the thermo-electric diagram. The path of the current will thus be represented by a line or path on the thermo-electric diagram. When the current flows in a single metal, A, from a point at a temperature OC to another at a temperature Oc, the path is represented by the line Aa, the points of which represent the state of the metal at intermediate temperatures. The form of the path depends on the nature of the metal and on the other influences which act on it besides temperature, such as stress and strain. Professor Tait, however, finds that for most of the metals except iron and nickel, the path on the thermo-electric diagram is a straight line.

When the current flows from the metal A to another metal B at the same temperature, the path is represented by AB, a vertical straight line. The circuit traversed by the electric current will thus be represented by a circuit on the thermo-electric diagram.

The heat generated while a unit of electricity moves along the path Aa is represented by the area of the figure $AaQPA$, bounded by the path Aa, the horizontal ordinate at a, the line of zero temperature and the horizontal ordinate at A. If this area lies on the right of the path, it represents heat generated; if it lies to the left of the path it represents heat absorbed.

179.] If electricity were a fluid, running through the conductor as water does through a tube, and always giving out or absorbing heat till its temperature is that of the conductor, then in passing from hot to cold it would give out heat and in passing from cold to hot it would absorb heat, and the amount of this heat would depend on the specific heat of the fluid.

In the diagram the specific heat of the fluid at A would be represented by the line aP, where a is the point where the tangent to the path at A cuts the line of zero temperature, and P is the intersection with the same line of the horizontal ordinate through A.

The line $Aa\alpha$ in the diagram is such that the electric entropy increases as the temperature rises. This is the case with copper, and therefore we may assert that the specific heat of electricity in copper is positive.

In other metals, as for instance iron, the electric entropy diminishes as the temperature rises, as is represented by the line βbB. The specific heat of electricity in such metals is negative, and at B is represented by the line βT.

180.] Thomson, who discovered first from theory and then by experimental verification the thermal effect of an electric current in an unequally heated metal, expresses the fact by saying that vitreous electricity carries heat with it in copper, while resinous electricity carries heat with it in iron.

We must remember, however, that these phrases are not intended by Thomson, and must not be understood by us, to imply that electricity either positive or negative is a fluid which can be heated or cooled and which has a definite specific heat. Since, therefore, the whole set of phrases are merely analogical we shall adhere to the ordinary convention according to which vitreous electricity is reckoned positive, and we shall say that the specific heat of electricity is positive in copper but negative in iron.

The obvious fact that no real fluid can have a negative specific heat need not disturb us, for we do not assert that electricity is a real fluid.

181.] Let us next consider a circuit consisting of two linear conductors of the metals A and B respectively, the two junctions being kept at different temperatures, represented in the diagram by OC and Oc. This electric circuit will be represented in the diagram by the circuit $AabBA$. If the current flows in the direction $AabB$ till one unit of electricity has been transmitted, the following thermal effects will take place.

(1) In the metal A heat will be generated as the electricity flows from the hot junction to the cold junction. The amount of this heat is represented bythe area $AaQPA$.

(2) At the cold junction, where the electricity passes from the metal A to the metal B, heat will be generated. The amount of this heat is represented by the area $abSQa$.

(3) In the metal B heat will be generated as the electricity flows from the cold junction to the hot junction. The amount of this heat is represented by the area $bBTSb$.

(4) At the hot junction, where the electricity passes from the metal B to the metal A, heat will be absorbed. The amount of this heat is represented by the area $BAPTB$. The reverse order of the letters shews that this area is to be taken negatively.

The whole heat generated is therefore represented by the area $AabBTPA$, and the whole heat absorbed by $BAPTB$. The total effect is therefore an absorption of heat represented by the area $AabBA$.

The energy corresponding to this heat cannot be lost. It is transformed into electrical work spent upon the current by an electromotive force acting in the direction of the current. Since the quantity of electricity transmitted by the current is supposed to be unity, the energy, which is the product of the electromotive force into the quantity of electricity transmitted, must be equal to the electromotive force itself.

Hence the electromotive force is represented by the area $AabBA$, and it acts in the direction represented by the order of the letters—that is,

Hot, metal A, cold, metal B, hot.

This electromotive force will, if the resistance of the circuit is finite, produce an actual current *. It was by means of such currents that the thermo-electric properties of metallic circuits were first discovered by Seebeck in 1822.

182.] The electrical effects due to heat were discovered before the thermal effects due to the electric current, but the application of the thermal effects of the current to determine the electromotive forces acting along different portions of the circuit is due to Sir W. Thomson †. It is manifest that in a heterogeneous circuit we cannot determine the electromotive force acting from the point A to the point B by simply connecting these points by wires to the electrodes of a galvanometer or electrometer, for we are ignorant of the electromotive forces acting at the junctions of these wires with the matter of the circuit at A and B.

But if we cause a current of known strength to flow from A to B, and if this current causes the generation of a quantity of heat equal to H in that portion of the circuit, and if no chemical, magnetic or other permanent effect takes place in the matter of the conductor between A and B, then we know that if Q is the total quantity of electricity which has been transmitted from A to B, and E the electromotive force in the direction from B to A which the current has to overcome, then the work done by the current is QE. This work is done within a definite region, namely the portion AB of the conductor, and it is entirely expended in generating heat within that region. Hence, if the quantity of heat generated in the portion AB is H, as expressed in dynamical measure, we have the equation

$$QE = H,$$

and since Q and H are capable of being measured we can determine the electromotive force E acting against the current. When the electromotive force acts in the same direction as the current is flowing, the quantity of heat generated is negative; or, in other words, there is an absorption of heat.

In this investigation we must remember that E represents the *whole* electromotive force acting against the current. Now part of this electromotive force arises from the electric resistance of the

* [The energy expended in driving the current will, if not otherwise employed, be ultimately converted into heat through the frictional resistance of the metals. The heat produced by this irreversible action must be distinguished from the Thomson and Peltier effects, and is represented on the Thermo-electric diagram by the area $ABbaA$.]

† *Trans. R. S. Edin.* 1854.

conductor. This part always acts against the current, and is proportional to the current according to Ohm's law.

The other part of the electromotive force acts in a definite direction, either from A to B or from B to A, and is independent of the direction of the current. It is generally this latter part of the electromotive force which is referred to as the electromotive force from A to B.

It is easy to eliminate the part due to resistance by making two experiments in which currents of equal strength are made to flow in one case from A to B and in the other from B to A. The excess of the heat generated in the second case over that generated in the first case, per unit of electricity transmitted, is numerically equal to twice the electromotive force from A to B.

183.] The total electromotive force round any circuit is easily measured by breaking the circuit in a place where it is homogeneous, and determining the difference of potentials of the two ends. This may be done by any of the ordinary methods for determining electromotive force or difference of potentials, because in this case the two ends are of the same substance and at the same temperature. But we cannot by this method determine how much of this electromotive force has its seat in a particular part of the circuit, as for instance, between A and B, where A and B are of different substances or at different temperatures. The only method by which we can determine where the electromotive force acts is that of measuring the heats generated or absorbed during the transmission of a unit of electricity from A to B.

184.] In the cases we have hitherto considered the only permanent effect of the current has been the generation or absorption of heat, for metals are not altered in any respect by the continuous flow of a current through them. But when the current flows from a metal to an electrolyte or from an electrolyte to a metal, there are chemical changes, and in applying the principle of the conservation of energy we must take account of these as well as of the thermal effects.

We shall consider the current as flowing through an electrolyte from the anode to the cathode. The fundamental phenomenon of electrolysis is the liberation of the components or ions of the electrolyte, the anion at the anode and the cation at the cathode. This is the only purely electrolytic effect; the subsequent phenomena depend on the nature of the ions, the electrodes and the electrolyte, and take place according to chemical and physical laws in a manner apparently independent of the electric current. Thus the ion, when liberated at the electrode, may behave in several different ways, according to the conditions in which it finds itself. It may be in such a condition that it acts neither on the electrode nor on the electrolyte, as when it is a gas which escapes in bubbles, or substance insoluble in the electrolyte, which is precipitated. It may be deposited on the surface of the electrode, as hydrogen is on platinum, and may adhere to it with various degrees of tenacity, from mere juxtaposition up to chemical combination. If it is soluble in the electrolyte, it will diffuse through the electrolyte according to the ordinary law of diffusion, and the rate of this diffusion is not, so far as we know, affected by the existence of the electric current through the electrolyte, for it is only when in combination, and not when in mere solution, that the current produces the electrolytic transfer of the ions. Thus when hydrogen is an ion, part of it may escape in bubbles, part of it may be condensed on the electrode, and part of it may be absorbed into the electrolyte without combination, and travel through it by ordinary diffusion.

185.] The liberated ion may also act chemically on the electrode or on the electrolyte. The results of such action are called secondary products of electrolysis, and these secondary products may remain at the surface of the electrodes, or may become diffused through the electrolyte. Thus, when the same current is passed, first through a solution of sulphate of soda between platinum electrodes, and then through sulphuric acid, equal volumes of oxygen are given off at the anodes of the two electrolytes, and equal volumes of hydrogen, each equal to double the volume of oxygen, are given off at the cathodes.

But if the electrolysis is conducted in suitable vessels, such as U-shaped tubes or vessels with a porous diaphragm, so that the substance surrounding each electrode may be examined, it is found that at the anode of the sulphate of soda there is an equivalent of sulphuric acid as well as an equivalent of oxygen, and at the cathode there is an equivalent of soda as well as two equivalents of hydrogen. It would at first sight appear as if (according to the old theory of the constitution of salts) the sulphate of soda were electrolysed into its constituents, sulphuric acid and soda, while the water of the solution is electrolysed at the same time into oxygen and hydrogen. But this explanation would involve the assumption that the same current which passing through dilute sulphuric acid electrolyses one equivalent of water, when it passes through solution of sulphate of soda electrolyses two equivalents, one of the salt and one of water, and this would be contrary to the law of electrochemical equivalents. But if we suppose that the components of sulphate of soda are not SO_3 and Na_2O, but SO_4 and Na_2—not sulphuric acid and soda but sulphion and sodium—then an equivalent of sulphion travels to the anode and is set free, but being unable to exist in a free state, it breaks up into sulphuric anhydride and oxygen, one equivalent of each. At the same time [two] equivalents of sodium are set free at the cathode, and then decompose the water of the solution, forming two equivalents of soda [$NaHO$] and two of hydrogen.

In the dilute sulphuric acid, the gases collected at the electrodes are the constituents of water, namely one volume of oxygen and two volumes of hydrogen. There is also an increase of sulphuric acid at the anode, but its amount is less than one equivalent.

186.] It follows from these considerations that in order to ascertain the electromotive force acting from a metal to an electrolyte, we must take account of the whole permanent effects of the passage of one unit of electricity from the metal to the electrolyte. Thus if the electrolyte is sulphate of zinc, with zinc electrodes, a certain amount of heat is generated at the anode for every unit of electricity and at the same time one equivalent of zinc combines with one equivalent of sulphion and forms sulphate of zinc. Now the quantity of heat generated when one equivalent of zinc combines with oxygen is known from the experiments of Andrews and others, and also the heat generated when an equivalent of oxide of zinc combines with sulphuric acid, and is dissolved in water so as to form a solution of sulphate of zinc of the same strength as that which surrounds the electrode. The sum of these quantities of heat, which we may call H, is equivalent to the total work done by the chemical action at the anode, which is therefore JH [where J represents Joule's equivalent or the mechanical equivalent of heat]. Let h be the quantity of heat generated at the anode during the passage of one unit of electricity, and let E be the electromotive force acting from the zinc to the electrolyte, that is, in the direction of the current. Then the work done in generating heat is Jh, and the work done in driving the current is E so that the equation of work is

$$JH = Jh + E$$

or

$$E = J(H - h).$$

Of these quantities H is known very accurately but it is somewhat difficult to measure h, the quantity of heat generated at the electrode, because the electrode must be in contact with the electrolyte, and therefore a large and unknown fraction of the heat generated will be carried away by conduction and convection through the electrolyte. The only method which seems likely to succeed is to compare the stationary temperature at a certain distance from the electrode with the temperature at the same point when in the place of the electrode we put a fine wire of known resistance through which we pass a known current so as to generate heat at a known rate. If the temperatures are equal in the two cases we may conclude that the heat is generated at the same rate in the zinc electrode and in the wire. But if the current is a strong one a very sensible portion of the whole heat generated will be due to the work done by the current in overcoming the ordinary resistance of the electrode and the electrolyte. As the electrode is generally made of a metal whose resistance is very small compared with that of the electrolyte, this frictional generation of heat will take place principally in the electrolyte. This frictional generation of heat may be made very small compared with the reversible part by diminishing the strength of the current, but then the rate of generation of heat becomes so small that it is difficult to measure it in the presence of unavoidable thermal disturbances, such as arise from changes in the temperature of the air, &c. The experimental investigation is therefore one of considerable difficulty and I am not aware that the electromotive force from a metal to an electrolyte has as yet been measured even approximately.* If, however, we assume that the electromotive forces from the metals A and B to the electrolyte C are A and B respectively, and that the thermo-electric powers of these metals at the temperature θ are a and b respectively, then the electromotive force from A to B at their junction is $(b-a)\,\theta$.

The total electromotive force round the circuit in the cyclical direction ABC is

$$(b-a)\,\theta+B-A.$$

On the Conservation of Energy in Electrolysis.

187.] Consider an electric current flowing in a circuit consisting partly of metals and partly of electrolytes placed in series.

During the passage of one unit of electricity through any section

* [See Art. 192 and last two paragraphs of note, p. 150.]

of the circuit one electrochemical equivalent of each of the electrolytes is electrolysed. There is therefore a definite amount of chemical action corresponding to a definite quantity of electricity passed through the circuit. The energy equivalent to any chemical process can be ascertained either directly or indirectly. When the process is such that it will go on of itself and if the only effect external to the system is the giving off of heat generated during the process, then the intrinsic energy of the system must be diminished during the process by a quantity of energy equivalent to the heat given out. If a material system consisting of definite quantities of so many chemical substances is capable of existing in several different states, and if the system will not of itself pass from one of these states (*A*) to another (*B*) we can still find the relative energy of the state (*A*) with respect to the state (*B*) provided we can cause both the state (*A*) and the state (*B*) to pass into the state (*C*) which we may suppose to be the state in which all the energies of combination of the system have been exhausted.

Thus if the substances in the system are oxygen, hydrogen and carbon and if the states (*A*) and (*B*) consist of two different hydrocarbons with free hydrogen and oxygen, we cannot in general cause the state (*A*) to pass into the state (*B*), but we can cause either (*A*) or (*B*) to pass into the state (*C*) in which all the hydrogen is combined with oxygen as water and all the carbon is combined with oxygen as carbonic acid. In this way the energy of the state (*A*) relatively to the state (*B*) can be determined by measurements of heat.

188.] It has been proved experimentally by Joule that the heat developed throughout the whole electric circuit is the same for the same amount of chemical action whatever be the resistance of the circuit provided no other form of energy than heat is given off by the system.

Thus in a battery the electrodes of which are connected by a short thick wire the current is very strong and the heat is generated principally in the cells of the battery and to a much smaller extent in the wire; but if the wire is long and thin, the heat generated in the wire is far greater than that generated in the cells, but if we take into account the heat generated in the wire as well as that generated in the cells, we find that the whole heat generated for each grain of zinc dissolved is the same in both cases.

189.] If, however, the circuit includes a cell in which dilute acid is electrolysed into oxygen and hydrogen the heat generated in the circuit per grain of zinc dissolved, is less than before, by the quantity of heat which would be generated if the oxygen and hydrogen evolved in the electrolytic cell were made to combine.

Or if the circuit includes an electromagnetic engine which is employed to do work, the heat generated in the circuit is less than that corresponding to the zinc consumed by an amount equal to the heat which would be generated if the work done by the engine were entirely expended in friction.

190.] If the arrangement is such that the amount of chemical action depends entirely on the quantity of electricity transmitted we can determine the electromotive force of the circuit by the following method, first given by Thomson (*Phil. Mag.*, Dec. 1851). Let the resistance of the circuit be made so great that the heat generated by the current in the electrolytes may be neglected. Let E be the electromotive force of the circuit; then the work done in driving one unit of electricity through the circuit is numerically equal to E. But during this process one electrochemical equivalent of the electrolyte undergoes the chemical process which goes on in the cell. Hence, if the energy given out during this process is entirely expended in maintaining the current, the dynamical value of the process must be numerically equal to E, the electromotive force of the circuit, or, as Thomson stated it,

'The electromotive force of an electrochemical apparatus is in absolute measure equal to the mechanical equivalent of the chemical action on one electrochemical equivalent of the substance.'

Examples.

191.] If the action in the cell consists in part of irreversible processes, such as

1. The frictional generation of heat by resistance in the electrolyte,
2. Diffusion of the primary or secondary products of electrolysis through the electrolyte, or,
3. Any other action which is not reversed when the direction of the current is reversed,

there will be a certain amount of dissipation of energy and the electromotive force of the circuit will be less than the loss of intrinsic energy corresponding to the electrolysis of one electro-chemical equivalent.

It is only the strictly reversible processes that must be taken into account in calculating the electromotive force of the circuit.

192.] The determination of the total electromotive force in an electrochemical circuit is therefore always possible. If, however, we wish to determine the precise points in the circuit where the different portions of this electromotive force are exerted, we find the investigation much more difficult than in the case of a purely metallic circuit.

For the chemical action at the junction of a metal with an electrolyte is generally of such a kind that it cannot take place by itself, that is to say, without an action equivalent to that which takes place at the other electrode. Thus, when a current passes between silver electrodes through fused chloride of silver, chlorine is liberated at the anode which immediately acts on the electrode so as to form chloride of silver and silver is deposited on the cathode.

Now we know the amount of heat given out when an equivalent of free chlorine combines with an equivalent of silver and this is equivalent to the energy which must be spent in electrolysing chloride of silver into free chlorine and free silver, but the process that takes place at the anode is the combination of silver, not with free chlorine, but with chlorine in the act of being electrolysed out of chloride of silver.*

* [The following note is an extract from Professor Maxwell's letter on Potential published in the *Electrician*, April 26th, 1879].

In a voltaic circuit the sum of the electromotive forces from zinc to the electrolyte, from the electrolyte to copper, and from copper to zinc, is not zero but is what is called the electromotive force of the circuit—a measurable quantity. Of these three electromotive forces only one can be separately measured by a legitimate process, that, namely, from copper to zinc.

Now it is found by thermoelectric experiments that this electromotive force is exceedingly small at ordinary temperatures, being less than a microvolt, and that it is from zinc to copper.

Hence the statement deduced from experiments in which air is the third medium, that the electromotive force from copper to zinc is ·75 volts, cannot be correct. In fact, what is really measured is the difference between the potential in air near the surface of copper, and the potential in air near the surface of zinc, the zinc and copper being in contact. The number ·75 is therefore the electromotive force, in volts of the circuit copper, zinc, air, copper, and is the sum of three electromotive forces, only one of which has as yet been measured.

Mr. J. Brown has shown (*Phil. Mag.* Aug. 1878, p. 142), by the divided ring method of Sir W. Thomson, that whereas copper is negative with respect to iron in air it is positive with respect to iron in hydrogen sulphide.

It would appear, therefore, that the reason why the results of the comparison of metals by the ordinary 'contact force' experiments harmonise so well with the comparison by dipping both metals in water or an oxidizing electrolyte is not because the electromotive force between a metal and a gas or an electrolyte is small, but because

On Constant Voltaic Elements.

193*.] When a series of experiments is made with a voltaic battery in which polarization occurs, the polarization diminishes during the time that the current is not flowing, so that when it begins to flow again the current is stronger than after it has flowed for some time. If, on the other hand, the resistance of the circuit is diminished by allowing the current to flow through a short shunt, then, when the current is again made to flow through the ordinary circuit, it is at first weaker than its normal strength on account of the great polarization produced by the use of the short circuit.

To get rid of these irregularities in the current, which are exceedingly troublesome in experiments involving exact measurements, it is necessary to get rid of the polarization, or at least to reduce it as much as possible.

It does not appear that there is much polarization at the surface of the zinc plate when immersed in a solution of sulphate of zinc or in dilute sulphuric acid. The principal seat of polarization is at the surface of the negative metal. When the fluid in which the negative metal is immersed is dilute sulphuric acid, it is seen to become covered with bubbles of hydrogen gas, arising from the electrolytic decomposition of the fluid. Of course these bubbles, by preventing the fluid from touching the metal, diminish the surface of contact and increase the resistance of the circuit. But besides the visible bubbles it is certain that there is a thin coating of hydrogen, probably not in a free state, adhering to the metal, and as we have seen that this coating is able to produce an electromotive force in the reverse direction, it must necessarily diminish the electromotive force of the battery.

Various plans have been adopted to get rid of this coating of hydrogen. It may be diminished to some extent by mechanical

the properties of air agree, to a certain extent, with those of oxidising electrolytes. For, if the active component of the electrolyte is sulphur, the results are quite different, and the same kind of difference occurs when hydrogen sulphide is substituted for air.

We know so little about the nature of the ions as they exist in an electrolyte that, even if we could measure the quantity of heat generated or absorbed when unit of electricity passes from a metal to an electrolyte, or from an electrolyte to a metal, we could not determine from this the value of the electromotive force from the metal to the electrolyte.

If this is the case with liquid electrolytes, we have still less hope of determining the electromotive force from a metal to a gas, for we cannot produce a current from the one to the other without tumultuary and non-reversible effects, such as disintegration of the metal and violent disturbance of the gas by the discontinuous discharge.

means, such as stirring the liquid, or rubbing the surface of the negative plate. In Smee's battery the negative plates are vertical, and covered with finely divided platinum from which the bubbles of hydrogen easily escape, and in their ascent produce a current of liquid which helps to brush off other bubbles as they are formed.

A far more efficacious method, however, is to employ chemical means. These are of two kinds. In the batteries of Grove and Bunsen the negative plate is immersed in a fluid rich in oxygen, and the hydrogen, instead of forming a coating on the plate, combines with this substance. In Grove's battery the plate is of platinum immersed in strong nitric acid. In Bunsen's first battery it is of carbon in the same acid. Chromic acid is also used for the same purpose, and has the advantage of being free from the acid fumes produced by the reduetion of nitric acid.

A different mode of getting rid of the hydrogen is by using copper as the negative metal, and covering the surface with a coat of oxide. This, however, rapidly disappears when it is used as the negative electrode. To renew it Joule has proposed to make the copper plates in the form of disks, half immersed in the liquid, and to rotate them slowly, so that the air may act on the parts exposed to it in turn.

The other method is by using as the liquid an electrolyte, the cation of which is a metal highly negative to zinc.

In Daniell's battery a copper plate is immersed in a saturated solution of sulphate of copper. When the current flows through the solution from the zinc to the copper no hydrogen appears on the copper plate, but copper is deposited on it. When the solution is saturated, and the current is not too strong, the copper appears to act as a true cation, the anion SO_4 travelling towards the zinc.

When these conditions are not fulfilled hydrogen is evolved at the cathode, but immediately acts on the solution, throwing down copper, and uniting with SO_4 to form oil of vitriol. When this is the case, the sulphate of copper next the copper plate is replaced by oil of vitriol, the liquid becomes colourless, and polarization by hydrogen gas again takes place. The copper deposited in this way is of a looser and more friable structure than that deposited by true electrolysis.

To ensure that the liquid in contact with the copper shall be saturated with sulphate of copper, crystals of this substance must be placed in the liquid close to the copper, so that when the solution is made weak by the deposition of the copper, more of the crystals may be dissolved.

We have seen that it is necessary that the liquid next the copper should be saturated with sulphate of copper. It is still more necessary that the liquid in which the zinc is immersed should be free from sulphate of copper. If any of this salt makes its way to the surface of the zinc it is reduced, and copper is deposited on the zinc. The zinc, copper, and fluid then form a little circuit in which rapid electrolytic action goes on, and the zinc is eaten away by an action which contributes nothing to the useful effect of the battery.

To prevent this, the zinc is immersed either in dilute sulphuric acid or in a solution of sulphate of zinc, and to prevent the solution of sulphate of copper from mixing with this liquid, the two liquids are separated by a division consisting of bladder or porous earthenware, which allows electrolysis to take place through it, but effectually prevents mixture of the fluids by visible currents.

In some batteries sawdust is used to prevent currents. The experiments of Graham, however, shew that the process of diffusion goes on nearly as rapidly when two liquids are separated by a division of this kind as when they are in direct contact, provided there are no visible currents, and it is probable that if a septum is employed which diminishes the diffusion, it will increase in exactly the same ratio the resistance of the element, because electrolytic conduction is a process the mathematical laws of which have the same form as those of diffusion, and whatever interferes with one must interfere equally with the other. The only difference is that diffusion is always going on, while the current flows only when the battery is in action.

In all forms of Daniell's battery the final result is that the sulphate of copper finds its way to the zinc and spoils the battery. To retard this result indefinitely, Sir W. Thomson* has constructed Daniell's battery in the form shewn in Fig. 41.

In each cell the copper plate is placed horizontally at the bottom, and a saturated solution of sulphate of zinc is poured over it. The zinc is in the form of a grating and is placed horizontally near the surface of the solution. A glass tube is placed vertically in the solution with its lower end just above the surface of the copper plate. Crystals of sulphate of copper are dropped down this tube, and, dissolving in the liquid, form a solution of greater density

* *Proc. R. S.*, Jan. 19, 1871.

than that of sulphate of zinc alone, so that it cannot get to the zinc except by diffusion. To retard this process of diffusion, a siphon, consisting of a glass tube stuffed with cotton wick, is placed with one extremity midway between the zinc and copper, and the other in a vessel outside the cell, so that the liquid is

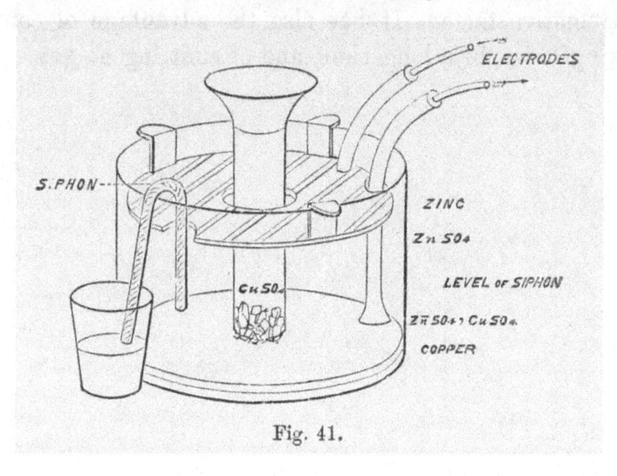

Fig. 41.

very slowly drawn off near the middle of its depth. To supply its place, water, or a weak solution of sulphate of zinc, is added above when required. In this way the greater part of the sulphate of copper rising through the liquid by diffusion is drawn off by the siphon before it reaches the zinc, and the zinc is surrounded by liquid nearly free from sulphate of copper, and having a very slow downward motion in the cell, which still further retards the upward motion of the sulphate of copper. During the action of the battery copper is deposited on the copper plate, and SO_4 travels slowly through the liquid to the zinc with which it combines, forming sulphate of zinc. Thus the liquid at the bottom becomes less dense by the deposition of the copper, and the liquid at the top becomes more dense by the addition of the zinc. To prevent this action from changing the order of density of the strata, and so producing instability and visible currents in the vessel, care must be taken to keep the tube well supplied with crystals of sulphate of copper, and to feed the cell above with a solution of sulphate of zinc sufficiently dilute to be lighter than any other stratum of the liquid in the cell.

Daniell's battery is by no means the most powerful in common use. The electromotive force of Grove's cell is 192,000,000, of Daniell's 107,900,000, and that of Bunsen's 188,000,000.

The resistance of Daniell's cell is in general greater than that of Grove's or Bunsen's of the same size.

These defects, however, are more than counterbalanced in all cases where exact measurements are required, by the fact that Daniell's cell exceeds every other known arrangement in constancy of electromotive force. It has also the advantage of continuing in working order for a long time, and of emitting no gas.

CHAPTER XI.

METHODS OF MAINTAINING AN ELECTRIC CURRENT.

194.] THE principal methods of maintaining a steady electric current are—

(1) The Frictional Machine.
(2) The Voltaic Battery.
(3) The Thermo-electric Battery.
(4) The Magneto-electric Machine.

(1) *The Frictional Electric Machine.*

195.] The electrification is here produced between the surfaces of two different substances, such as glass and amalgam or ebonite and fur. By the motion of the machine one of these electrified surfaces is carried away from the other, and both are made to discharge their electrification into the electrodes of the machine, from which the current is conveyed along any required circuit.

In the ordinary form of the machine a circular plate or a cylinder of glass is made to revolve about its axis. Let us suppose that the revolving part is a plate of glass. The rubber is fixed so that it presses against the surface of the plate as it rotates. The surface of the rubber is of leather, on which is spread an amalgam of zinc and mercury. By the friction between the glass and the amalgam the surface of the glass becomes electrified positively, and that of the rubber negatively. As the plate revolves the electrified surface of the glass is carried away from under the rubber, and another part of the surface of the glass, previously unelectrified, is brought under the rubber to be electrified. As long as the oppositely electrified surfaces of the glass and the rubber remain in contact, the electrical effects in the neighbourhood are very small, but when the glass is removed from the rubber, strong electrical forces are developed. The potential of the rubber becomes negative, and as, on account of the amalgam, it conducts freely its electrification is at once carried off to the negative electrode. At the same time the potential of the electrified glass becomes highly positive, but as the glass is an insulating substance it does not so readily part with its electrification. The positive electrode of the machine is therefore furnished with a comb, consisting of a number of sharp pointed wires terminating near the electrified surface of the glass. As the potential at the surface of the glass is much higher than that of the comb there is a great accumulation of negative electrification at the point of the comb, and this breaks into a negative electric glow accompanied by an electric wind blowing from the comb to the glass. The negatively electrified particles of air spread themselves over the positively electrified surface of the glass, and cause the electrification of the glass to be discharged. It is possible, however, that part of them may be carried round with the glass till they are wiped off by the rubber, though I have not been able to obtain experimental evidence of this.

Thus the rotation of the machine carries the positive electrification of the surface of the glass from the rubber to the comb, and the negative electric wind of the comb either neutralizes the positively electrified surface, or is carried round with it to the rubber, so that there is a continual current of positive electricity kept up from the rubber to the comb, or, what is the same thing, of negative electricity from the comb to the rubber, or, since the mode of expressing the fact is indifferent, we may, if we please, describe it as consisting of a positive current in the one direction combined with a negative current in the other the arithmetical sum of these two imaginary currents being the actual current observed. The action of the machine thus depends on the electrification of the surface of the glass by the rubber, the convection of this electrification, by the motion of the machine, to the comb and the discharge of the electrification by the comb.

196.] The strength of the current produced depends on the surface-density of the electrification, the area of the electrified surface and the number of turns in a minute.

The electromotive force of the machine is the excess of the potential of the comb above that of the rubber. The most convenient test of the electromotive force of an electrical machine is the length of the sparks which it will give.

During the passage of the electrified surface from the rubber to the comb it is passing from places of low to places of high potential, and is therefore acted on by a force in the direction opposite to that of its motion. The work done in turning the machine therefore exceeds that necessary to overcome the friction of the rubber, the axle, and other mechanical resistances by the electrical work done in carrying the electricity from the rubber to the comb.

At every point of its course the electricity on the surface of the glass plate is acted on by a force the value of which is measured by the rate at which the potential varies from one point to another of the surface. If this force exceeds a certain value it will cause the electrification to slide along the surface of the plate, and this will take place under the action of a much smaller force than that which is required to remove the electricity from the surface. This discharge along the surface of the plate may be seen when the electric machine is worked in a dark room, and it is evident that the electricity which thus flashes back is so much lost from the principal current of the machine.

In order that the machine may work to the best advantage this slipping back of the electricity must be prevented. The slipping takes place whenever the rate of variation of the potential from point to point of the surface exceeds a certain value. If by any distribution of the electrification the rate of variation of the potential can be kept just below this value all the way from the rubber to the comb the electromotive force of the machine will have its highest possible value.

In most electrical machines flaps of oiled silk are attached to the rubber so that as the plate revolves the electrified surface as it leaves the rubber is covered with the silk flap which extends from the rubber nearly up to the comb. These silk flaps become negatively electrified and therefore adhere of themselves to the surface of the glass. If in any part of the revolution of the plate, the rate of increase of the potential is so great that a slipping back of the electrification occurs, the positive electricity which so slips back neutralizes part of the negative electrification of the silk flap and so raises the electric potential just behind the place where the slipping occurred. In this way the slope of the electric potential is equalized and the electromotive force of the machine is raised to its highest possible value, so as to give the longest sparks which a machine of given dimensions can furnish.

When the silk flaps are removed the slope of the potential

becomes much greater close to the rubber than at any other place, the electricity slips back on the glass just as it leaves the rubber and very little electricity and that at a comparatively low potential reaches the comb.

In the best machines, in which the slope of the potential is uniform from the rubber to the comb, the length of the spark must depend principally on the distance between the rubber and the comb. Hence a machine which, like Winter's, has the rubber and the comb at opposite extremities of a diameter of the plate will give a longer spark than one from a machine whose plate has the same diameter but which like Cuthbertson's has two rubbers and two combs, the distance between each rubber and its comb being a quadrant.

On Machines producing Electrification by Mechanical Work.

197*.] In the ordinary frictional electrical machine the work done in overcoming friction is far greater than that done in increasing the electrification. Hence any arrangement by which the electrification may be produced entirely by mechanical work against the electrical forces is of scientific importance if not of practical value. The first machine of this kind seems to have been Nicholson's Revolving Doubler, described in the *Philosophical Transactions* for 1788 as 'an instrument which by the turning of a Winch produces the two states of Electricity without friction or communication with the Earth.'

198*.] It was by means of the revolving doubler that Volta succeeded in developing from the electrification of the pile an electrification capable of affecting his electrometer. Instruments on the same principle have been invented independently by Mr. C. F. Varley *, and Sir W. Thomson.

These instruments consist essentially of insulated conductors of various forms, some fixed and others moveable. The moveable conductors are called Carriers, and the fixed ones may be called Inductors, Receivers, and Regenerators. The inductors and receivers are so formed that when the carriers arrive at certain points in their revolution they are almost completely surrounded by a conducting body. As the inductors and receivers cannot completely surround the carrier and at the same time allow it to move freely in and out without a complicated arrangement of moveable pieces, the instrument is not theoretically perfect without a pair of re-

* Specification of Patent, Jan. 27, 1860, No. 206.

generators, which store up the small amount of electricity which the carriers retain when they emerge from the receivers.

For the present, however, we may suppose the inductors and receivers to surround the carrier completely when it is within them, in which case the theory is much simplified.

We shall suppose the machine to consist of two inductors A and C, and of two receivers B and D, with two carriers F and G.

Suppose the inductor A to be positively electrified so that its potential is A, and that the carrier F is within it and is at potential F. Then, if Q is the coefficient of induction (taken positive) between A and F, the quantity of electricity on the carrier will be $Q(F-A)$.

If the carrier, while within the inductor, is put in connexion with the earth, then $F=0$, and the charge on the carrier will be $-QA$, a negative quantity. Let the carrier be carried round till it is within the receiver B, and let it then come in contact with a spring so as to be in electrical connexion with B. It will then, as was shewn in Art. 20, become completely discharged, and will communicate its whole negative charge to the receiver B.

The carrier will next enter the inductor C, which we shall suppose charged negatively. While within C it is put in connexion with the earth and thus acquires a positive charge, which it carries off and communicates to the receiver D, and so on.

In this way, if the potentials of the inductors remain always constant, the receivers B and D receive successive charges, which are the same for every revolution of the carrier, and thus every revolution produces an equal increment of electricity in the receivers.

But by putting the inductor A in communication with the receiver D, and the inductor C with the receiver B, the potentials of the inductors will be continually increased, and the quantity of electricity communicated to the receivers in each revolution will continually increase.

For instance, let the potential of A and D be U, and that of B and C, V, and when the carrier is within A let the charge on A and C be x, and that on the carrier z, then, since the potential of the carrier is zero, being in contact with earth, its charge is $z=-QU$. The carrier enters B with this charge and communicates it to B. If the capacity of B and C is B, their potential will be changed from V to $V-\frac{Q}{B}U$.

If the other carrier has at the same time carried a charge $-QV$ from C to D, it will change the potential of A and O from U to $U-\frac{Q'}{A}V$, if Q' is the coefficient of induction between the carrier and C, and A the capacity of A and D. If, therefore, U_n and V_n be the potentials of the two inductors after n half revolutions, and U_{n+1} and V_{n+1} after $n+1$ half revolutions,

$$U_{n+1}=U_n-\frac{Q'}{A}V_n,$$

$$V_{n+1}=V_n-\frac{Q}{B}U_n.$$

If we write $p^2=\frac{Q}{B}$ and $q^2=\frac{Q'}{A}$, we find

$$pU_{n+1}+qV_{n+1}=(pU_n+qV_n)(1-pq)=(pU_0+qV_0)(1-pq)^{n+1},$$
$$pU_{n+1}-qV_{n+1}=(pU_n-qV_n)(1+pq)=(pU_0-qV_0)(1+pq)^{n+1}.$$

Hence

$$U_n=U_0((1-pq)^n+(1+pq)^n)+\frac{q}{p}V_0((1-pq)^n-(1+pq)^n),$$

$$V_n=\frac{p}{q}U_0((1-pq)^n-(1+pq)^n)+V_0((1-pq)^n+(1+pq)^n).$$

It appears from these equations that the quantity $pU+qV$ continually diminishes, so that whatever be the initial state of electrification the receivers are ultimately oppositely electrified, so that the potentials of A and B are in the ratio of q to $-p$.

On the other hand, the quantity $pU-qV$ continually increases, so that, however little pU may exceed or fall short of qV at first, the difference will be increased in a geometrical ratio in each revolution till the electromotive forces become so great that the insulation of the apparatus is overcome.

Instruments of this kind may be used for various purposes.

For producing a copious supply of electricity at a high potential, as is done by means of Mr. Varley's large machine.

For adjusting the charge of a condenser, as in the case of Thomson's electrometer, the charge of which can be increased or diminished by a few turns of a very small machine of this kind, which is then called a Replenisher.

For multiplying small differences of potential. The inductors may be charged at first to an exceedingly small potential, as, for instance, that due to a thermo-electric pair, then, by turning the machine, the difference of potentials may be continually multiplied till it becomes capable of measurement by an ordinary electrometer. By determining by experiment the ratio of increase of this difference due to each turn of the machine, the original electromotive force with which the inductors were charged may be deduced from the number of turns and the final electrification.

In most of these instruments the carriers are made to revolve about an axis and to come into the proper positions with respect to the inductors by turning an axle. The connexions are made by means of springs so placed that the carriers come in contact with them at the proper instants.

199*.] Sir W. Thomson*, however, has constructed a machine for multiplying electrical charges in which the carriers are drops of water falling out of the inside of an inductor into an insulated receiver. The receiver is thus continually supplied with electricity of opposite sign to that of the inductor. If the inductor is electrified positively, the receiver will receive a continually increasing charge of negative electricity.

The water is made to escape from the receiver by means of a funnel, the nozzle of which is almost surrounded by the metal of the receiver. The drops falling from this nozzle are therefore nearly free from electrification. Another inductor and receiver of the same construction are arranged so that the inductor of the one system is in connexion with the receiver of the other. The rate of increase of charge of the receivers is thus no longer constant, but increases in a geometrical progression with the time, the charges of the two receivers being of opposite signs. This increase goes on till the falling drops are so diverted from their course by the electrical action that they fall outside of the receiver or even strike the inductor.

In this instrument the energy of the electrification is drawn from that of the falling drops.

200.] In Holtz's 'Influence-Machine' a plate of varnished glass is made to rotate in front of a fixed plate of varnished glass. The inductors consist of two pointed pieces of card sometimes covered with tin foil and placed on the further side of the fixed plate so that their points are at opposite extremities of a diameter. Holes are cut in the fixed plate opposite the points of the inductors. The electrodes are first put in connexion with each other and the machine is set in rotation. One of the inductors is then electrified, either by an ordinary machine or by an excited piece of ebonite.

* *Proc. R. S.*, June 20, 1867.

Let us suppose it electrified positively. The comb in front of the charged inductor immediately begins to glow and discharges negative electricity against the rotating disk. This negative electrification is carried round by the disk to the other side where it is free from the influence of the positive inductor. The other inductor now discharges positive electricity from its point and becomes itself negatively charged, and the comb of the negative electrode discharges positive electricity, which is carried round the disk on the other side back to the positive electrode. In this way there is kept up an electric current from the positive to the negative electrode. A rushing noise is heard and in the dark a glow is seen extending itself from the positive comb over the surface of the rotating disk in the direction opposite to its motion. If the electrodes are now gradually separated a succession of sparks will pass between them.

Influence Machine.

1865. Holtz exhibited his machine to the Berlin Academy, April 1865. 8 to 10 cm. diam.

1866. Töpler, metal inductors, two metal carriers on a glass disk.

1867. Töpler's multiple machine, 8 rotating disks, 32 cm. diam. sparks 6 to 9 cm.

1867. Holtz with two disks rotating oppositely.

1868. Kundt.

Carré, inductor disk 38 cm. induced 49, spark 15 to 18.

201*.] In the electrical machines already described sparks occur whenever the carrier comes in contact with a conductor at a different potential from its own.

Now we have shewn that whenever this occurs there is a loss of energy, and therefore the whole work employed in turning the machine is not converted into electrification in an available form, but part is spent in producing the heat and noise of electric sparks.

I have therefore thought it desirable to shew how an electrical machine may be constructed which is not subject to this loss of efficiency. I do not propose it as a useful form of machine, but as an example of the method by which the contrivance called in heat-engines a regenerator may be applied to an electrical machine to prevent loss of work.

In the figure let A, B, C, A', B', C' represent hollow fixed conductors, so arranged that the carrier P passes in succession within each of them. Of these A, A' and B, B' nearly surround the carrier when it is at the middle point of its passage, but C, C' do not cover it so much.

We shall suppose A, B, C to be connected with a Leyden jar of great capacity at potential V, and A', B', C' to be connected with another jar at potential $-V'$.

P is one of the carriers moving in a circle from A to C', &c., and touching in its course certain springs, of which a and a' are connected with A and A' respectively, and e, e' are connected with the earth.

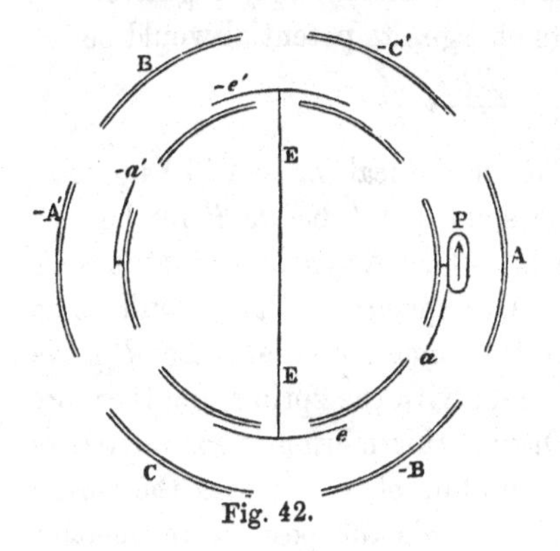

Fig. 42.

Let us suppose that when the carrier P is in the middle of A the coefficient of induction between P and A is $-A$. The capacity of P in this position is greater than A, since it is not completely surrounded by the receiver A. Let it be $A+a$.

Then if the potential of P is U, and that of A, V, the charge on P will be $(A+a)U-AV$.

Now let P be in contact with the spring a when in the middle of the receiver A, then the potential of P is V, the same as that of A, and its charge is therefore aV.

If P now leaves the spring a it carries with it the charge aV. As P leaves A its potential diminishes, and it diminishes still more when it comes within the influence of C', which is negatively electrified.

If when P comes within C its coefficient of induction on C is $-C'$, and its capacity is $C'+c'$, then, if U is the potential of P the charge on P is

$$(C'+c')U+C'V'=aV.$$

If

$$C'V'=aV,$$

then at this point U the potential of P will be reduced to zero.

Let P at this point come in contact with the spring e' which is connected with the earth. Since the potential of P is equal to that of the spring there will be no spark at contact.

This conductor C', by which the carrier is enabled to be connected to earth without a spark, answers to the contrivance called a regenerator in heat-engines. We shall therefore call it a Regenerator.

Now let P move on, still in contact with the earth-spring e', till it comes into the middle of the inductor B, the potential of which is V. If $-B$ is the coefficient of induction between P and B at this point, then, since $U=0$ the charge on P will be $-BV$.

When P moves away from the earth-spring it carries this charge with it. As it moves out of the positive inductor B towards the negative receiver A' its potential will be increasingly negative. At the middle of A', if it retained its charge, its potential would be

$$-\frac{A'V'+BV}{A'+a'},$$

and if BV is greater than $a'V'$ its numerical value will be greater than that of V'. Hence there is some point before P reaches the middle of A' where its potential is $-V'$. At this point let it come in contact with the negative receiver-spring a'. There will be no spark since the two bodies are at the same potential. Let P move on to the middle of A', still in contact with the spring, and therefore at the same potential with A'. During this motion it communicates a negative charge to A'. At the middle of A' it leaves the spring and carries away a charge $-a'V'$ towards the positive regenerator C, where its potential is reduced to zero and it touches the earth-spring e. It then slides along the earth-spring into the negative inductor B', during which motion it acquires a positive charge $B'V'$ which it finally communicates to the positive receiver A, and the cycle of operations is repeated.

During this cycle the positive receiver has lost a charge aV and gained a charge $B'V'$. Hence the total gain of positive electricity is

$$B'V'-aV.$$

Similarly the total gain of negative electricity is $BV-a'V'$.

By making the inductors so as to be as close to the surface of the carrier as is consistent with insulation, B and B' may be made large, and by making the receivers so as nearly to surround the carrier when it is within them, a and a' may be made very small, and then the charges of both the Leyden jars will be increased in every revolution.

The conditions to be fulfilled by the regenerators are

$$C'V'=aV, \quad \text{and} \quad CV=a'V'.$$

Since a and a' are small the regenerators do not require to be either large or very close to the carriers.

Coulomb's Torsion Balance.

202*.] A great number of the experiments by which Coulomb established the fundamental laws of electricity were made by measuring the force between two small spheres charged with electricity, one of which was fixed while the other was held in equilibrium by two forces, the electrical action between the spheres, and the torsional elasticity of a glass fibre or metal wire.

The balance of torsion consists of a horizontal arm of gum-lac, suspended by a fine wire or glass fibre, and carrying at one end a little sphere of elder pith, smoothly gilt. The suspension wire is fastened above to the vertical axis of an arm which can be moved round a horizontal graduated circle, so as to twist the upper end of the wire about its own axis any number of degrees.

The whole of this apparatus is enclosed in a case. Another little sphere is so mounted on an insulating stem that it can be charged and introduced into the case through a hole, and brought so that its centre coincides with a definite point in the horizontal circle described by the suspended sphere. The position of the suspended sphere is ascertained by means of a graduated circle engraved on the cylindrical glass case of the instrument.

Now suppose both spheres charged, and the suspended sphere in equilibrium in a known position such that the torsion-arm makes an angle θ with the radius through the centre of the fixed sphere. The distance of the centres is then $2a\sin\frac{1}{2}\theta$, where a is the radius of the torsion-arm, and if F is the force between the spheres the moment of this force about the axis of torsion is $Fa\cos\frac{1}{2}\theta$.

Let both spheres be completely discharged, and let the torsion-arm now be in equilibrium at an angle ϕ with the radius through the fixed sphere.

Then the angle through which the electrical force twisted the torsion-arm must have been $\theta-\phi$, and if M is the moment of the torsional elasticity of the fibre, we shall have the equation

$$Fa\cos\tfrac{1}{2}\theta = M(\theta-\phi).$$

Hence, if we can ascertain M, we can determine F, the actual force between the spheres at the distance $2a\sin\frac{1}{2}\theta$.

To find M, the moment of torsion, let I be the moment of inertia of the torsion-arm, and T the time of a double vibration of the arm under the action of the torsional elasticity, then

$$M = \frac{1}{4\pi^2} IT^2.$$

In all electrometers it is of the greatest importance to know what force we are measuring. The force acting on the suspended sphere is due partly to the direct action of the fixed sphere, but partly also to the electrification, if any, of the sides of the case.

If the case is made of glass it is impossible to determine the electrification of its surface otherwise than by very difficult measurements at every point. If, however, either the case is made of metal, or if a metallic case which almost completely encloses the apparatus is placed as a screen between the spheres and the glass case, the electrification of the inside of the metal screen will depend entirely on that of the spheres, and the electrification of the glass case will have no influence on the spheres. In this way we may avoid any indefiniteness due to the action of the case.

To illustrate this by an example in which we can calculate all the effects, let us suppose that the case is a sphere of radius b, that the centre of motion of the torsion-arm coincides with the centre of the sphere and that its radius is a; that the charges on the two spheres are E_1 and E_2, and that the angle between their positions is θ; that the fixed sphere is at a distance a_1 from the centre, and that r is the distance between the two small spheres.

Neglecting for the present the effect of induction on the distribution of electricity on the small spheres, the force between them will be a repulsion

$$= \frac{EE_1}{r^2},$$

and the moment of this force round a vertical axis through the centre will be

$$\frac{EE_1 aa_1 \sin\theta}{r^3}.$$

The image of E_1 due to the spherical surface of the case is a point in the same radius at a distance $\frac{b^2}{a_1}$ with a charge $-E_1\frac{b}{a_1}$, and the moment of the attraction between E and this image about the axis of suspension is

$$EE_1\frac{b}{a_1}\frac{a\dfrac{b^2}{a_1}\sin\theta}{\left\{a^2-2\dfrac{ab^2}{a_1}\cos\theta+\dfrac{b^4}{a_1{}^2}\right\}^{\frac{3}{2}}}$$

$$= EE_1\frac{aa_1\sin\theta}{b^3\left\{1-2\dfrac{aa_1}{b^2}\cos\theta+\dfrac{a^2a^2{}_1}{b^4}\right\}^{\frac{3}{2}}}.$$

If b, the radius of the spherical case, is large compared with a and a_1, the distances of the spheres from the centre, we may neglect the second and third terms of the factor in the denominator. The whole moment tending to turn the torsion-arm may then be written

$$EE_1 a a_1 \sin\theta \left\{\frac{1}{r^3} - \frac{1}{b^3}\right\} = M(\theta - \phi).$$

Electrometers for the Measurement of Potentials.

203*.] In all electrometers the moveable part is a body charged with electricity, and its potential is different from that of certain of the fixed parts round it. When, as in Coulomb's method, an insulated body having a certain charge is used, it is the charge which is the direct object of measurement. We may, however, connect the balls of Coulomb's electrometer, by means of fine wires, with different conductors. The charges of the balls will then depend on the values of the potentials of these conductors and on the potential of the case of the instrument. The charge on each ball will be approximately equal to its radius multiplied by the excess of its potential over that of the case of the instrument, provided the radii of the balls are small compared with their distances from each other and from the sides or opening of the case.

Coulomb's form of apparatus, however, is not well adapted for measurements of this kind, owing to the smallness of the force between spheres at the proper distances when the difference of potentials is small. A more convenient form is that of the Attracted Disk Electrometer. The first electrometers on this principle were constructed by Sir W. Snow Harris *. They have since been brought to great perfection, both in theory and construction, by Sir W. Thomson †.

When two disks at different potentials are brought face to face with a small interval between them there will be a nearly uniform electrification on the opposite faces and very little electrification on the backs of the disks, provided there are no other conductors or electrified bodies in the neighbourhood. The charge on the positive disk will be approximately proportional to its area, and to the difference of potentials of the disks, and inversely as the distance

* *Phil. Trans.* 1834.

† See an excellent report on Electrometers by Sir W. Thomson. *Report of the British Association*, Dundee, 1867.

between them. Hence, by making the areas of the disks large and the distance between them small, a small difference of potential may give rise to a measurable force of attraction.

204*.] The addition of the guard-ring to the attracted disk is one of the chief improvements which Sir W. Thomson has made on the apparatus.

Instead of suspending the whole of one of the disks and determining the force acting upon it, a central portion of the disk is separated from the rest to form the attracted disk, and the outer ring forming the remainder of the disk is fixed. In this way the

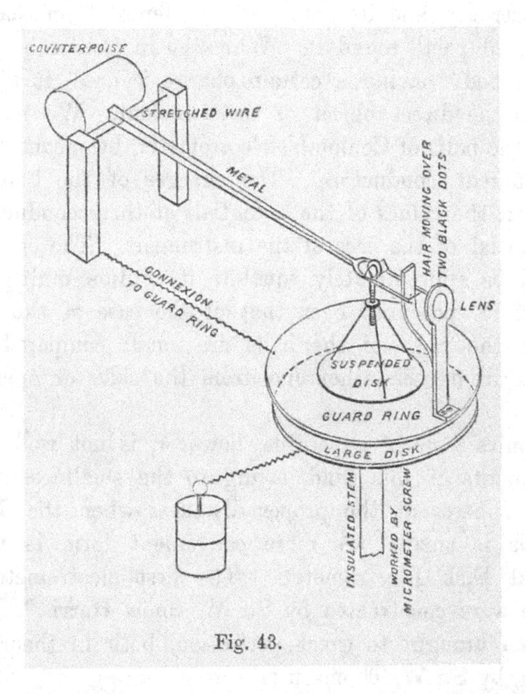

Fig. 43.

force is measured only on that part of the disk where it is most regular, and the want of uniformity of the electrification near the edge is of no importance, as it occurs on the guard-ring and not on the suspended part of the disk.

Besides this, by connecting the guard-ring with a metal case surrounding the back of the attracted disk and all its suspending apparatus, the electrification of the back of the disk is rendered impossible, for it is part of the inner surface of a closed hollow conductor all at the same potential.

Thomson's Absolute Electrometer therefore consists essentially of two parallel plates at different potentials, one of which is made so that a certain area, no part of which is near the edge of the plate, is moveable under the action of electric force. To fix our ideas we may suppose the attracted disk and guard-ring uppermost. The fixed disk is horizontal, and is mounted on an insulating stem which has a measurable vertical motion given to it by means of a micrometer screw. The guard-ring is at least as large as the fixed disk; its lower surface is truly plane and parallel to the fixed disk. A delicate balance is erected on the guard-ring to which is suspended a light moveable disk which almost fills the circular aperture in the guard-ring without rubbing against its sides. The lower surface of the suspended disk must be truly plane, and we must have the means of knowing when its plane coincides with that of the lower surface of the guard-ring, so as to form a single plane interrupted only by the narrow interval between the disk and its guard-ring.

For this purpose the lower disk is screwed up till it is in contact with the guard-ring, and the suspended disk is allowed to rest upon the lower disk, so that its lower surface is in the same plane as that of the guard-ring. Its position with respect to the guard-ring is then ascertained by means of a system of fiducial marks. Sir W. Thomson generally uses for this purpose a black hair attached to the moveable part. This hair moves up or down just in front of two black dots on a white enamelled ground and is viewed along with these dots by means of a plano convex lens with the plane side next the eye. If the hair as seen through the lens appears straight and bisects the interval between the black dots it is said to be in its *sighted position*, and indicates that the suspended disk with which it moves is in its proper position as regards height. The horizontality of the suspended disk may be tested by comparing the reflexion of part of any object from its upper surface with that of the remainder of the same object from the upper surface of the guard-ring.

The balance is then arranged so that when a known weight is placed on the centre of the suspended disk it is in equilibrium in its sighted position, the whole apparatus being freed from electrification by putting every part in metallic communication. A metal case is placed over the guard-ring so as to enclose the balance and suspended disk, sufficient apertures being left to see the fiducial marks.

The guard-ring, case, and suspended disk are all in metallic communication with each other, but are insulated from the other parts of the apparatus.

Now let it be required to measure the difference of potentials of two conductors. The conductors are put in communication with the upper and lower disks respectively by means of wires, the weight is taken off the suspended disk, and the lower disk is moved up by means of the micrometer screw till the electrical attraction brings the suspended disk down to its sighted position. We then know that the attraction between the disks is equal to the weight which brought the disk to its sighted position.

If W be the numerical value of the weight, and g the force of gravity, the force is Wg, and if A is the area of the suspended disk, D the distance between the disks, and V the difference of the potentials of the disks,

$$Wg = \frac{V^2 A}{8\pi D^2},$$

$$\text{or} \quad V = D\sqrt{\frac{8\pi g W}{A}}.$$

If the suspended disk is circular, of radius R, and if the radius of the aperture of the guard-ring is R', then

$$A = \tfrac{1}{2}\pi(R^2 + R'^2)^*, \text{ and } V = 4D\sqrt{\frac{gW}{R^2 + R'^2}}.$$

* Let us denote the radius of the suspended disk by R, and that of the aperture of the guard-ring by R', then the breadth of the annular interval between the disk and the ring will be $B = R' - R$.

If the distance between the suspended disk and the large fixed disk is D, and the difference of potentials between these disks is V, then (see *Electricity and Magnetism*, Art. 201) the quantity of electricity on the suspended disk will be

$$Q = V\left\{\frac{R^2 + R'^2}{8D} - \frac{R'^2 - R^2}{8D}\frac{a}{D + a}\right\},$$

$$\text{where} \quad a = B\frac{\log_e 2}{\pi}, \quad \text{or} \quad a = 0.220635\,(R' - R).$$

If the surface of the guard-ring is not exactly in the plane of the surface of the suspended disk, let us suppose that the distance between the fixed disk and the guard ring is not D but $D + z = D'$, then (see *Electricity and Magnetism*, Art. 205) there will be an additional charge of electricity near the edge of the disk on account of its height z above the general surface of the guard-ring. The whole charge in this case is therefore

$$Q = V\left\{\frac{R^2 + R'^2}{8D} - \frac{R'^2 - R^2}{8D}\frac{a}{D + a} + \frac{R + R'}{D}(D - D)\log_e\frac{4\pi(R + R')}{D' - D}\right\},$$

and in the expression for the attraction we must substitute for A, the area of the disk, the corrected quantity

$$A = \tfrac{1}{2}\pi\left\{R^2 + R'^2 - (R'^2 - R^2)\frac{a}{D + a} + 8(R + R')(D' - D)\log_e\frac{4\pi(R + R')}{D' - D}\right\},$$

205*.] Since there is always some uncertainty in determining the micrometer reading corresponding to $D = 0$, and since any error in the position of the suspended disk is most important when D is small, Sir W. Thomson prefers to make all his measurements depend on differences of the electromotive force V. Thus, if V and V' are two potentials, and D and D' the corresponding distances,

$$V - V' = (D - D')\sqrt{\frac{8\pi g W}{A}}.$$

For instance, in order to measure the electromotive force of a galvanic battery, two electrometers are used.

By means of a condenser, kept charged if necessary by a replenisher, the lower disk of the principal electrometer is maintained at a constant potential. This is tested by connecting the lower disk of the principal electrometer with the lower disk of a secondary electrometer, the suspended disk of which is connected with the earth. The distance between the disks of the secondary electrometer and the force required to bring the suspended disk to its sighted position being constant, if we raise the potential of the condenser till the secondary electrometer is in its sighted position, we know that the potential of the lower disk of the principal electrometer exceeds that of the earth by a constant quantity which we may call V.

If we now connect the positive electrode of the battery to earth, and connect the suspended disk of the principal electrometer to the negative electrode, the difference of potentials between the disks will be $V + v$, if v is the electromotive force of the battery. Let D be the reading of the micrometer in this case, and let D' be the reading when the suspended disk is connected with earth, then

$$v = (D - D')\sqrt{\frac{8\pi g W}{A}}.$$

In this way a small electromotive force v may be measured by the electrometer with the disks at conveniently measurable distances. When the distance is too small a small change of absolute distance makes a great change in the force, since the

where R = radius of suspended disk,
R' = radius of aperture in the guard-ring,
D = distance between fixed and suspended disks,
D' = distance between fixed disk and guard-ring,
$\alpha = 0.220635\,(R' - R)$.

When α is small compared with D we may neglect the second term, and when $D' - D$ is small we may neglect the last term.

force varies inversely as the square of the distance, so that any error in the absolute distance introduces a large error in the result unless the distance is large compared with the limits of error of the micrometer screw.

The effect of small irregularities of form in the surfaces of the disks and of the interval between them diminish according to the inverse cube and higher inverse powers of the distance, and whatever be the form of a corrugated surface, the eminences of which just reach a plane surface, the electrical effect at any distance which is considerable compared to the breadth of the corrugations, is the same as that of a plane at a certain small distance behind the plane of the tops of the eminences.

By means of the auxiliary electrification, tested by the auxiliary electrometer, a proper interval between the disks is secured.

The auxiliary electrometer may be of a simpler construction, in which there is no provision for the determination of the force of attraction in absolute measure, since all that is wanted is to secure a constant electrification. Such an electrometer may be called a *gauge* electrometer.

This method of using an auxiliary electrification besides the electrification to be measured is called the Heterostatic method of electrometry, in opposition to the Idiostatic method in which the whole effect is produced by the electrification to be measured.

In several forms of the attracted disk electrometer, the attracted disk is placed at one end of an arm which is supported by being attached to a platinum wire passing through its centre of gravity and kept stretched by means of a spring. The other end of the arm carries the hair which is brought to a sighted position by altering the distance between the disks, and so adjusting the force of the electric attraction to a constant value. In these electrometers this force is not in general determined in absolute measure, but is known to be constant, provided the torsional elasticity of the platinum wire does not change.

The whole apparatus is placed in a Leyden jar, of which the inner surface is charged and connected with the attracted disk and guard-ring. The other disk is worked by a micrometer screw and is connected first with the earth and then with the conductor whose potential is to be measured. The difference of readings multiplied by a constant to be determined for each electrometer gives the potential required.

On the Measurement of Electric Potential.

206*.] In order to determine large differences of potential in absolute measure we may employ the attracted disk electrometer, and compare the attraction with the effect of a weight. If at the same time we measure the difference of potential of the same conductors by means of the quadrant electrometer, we shall ascertain the absolute value of certain readings of the scale of the quadrant electrometer, and in this way we may deduce the value of the scale readings of the quadrant electrometer in terms of the potential of the suspended part, and the moment of torsion of the suspension apparatus.

To ascertain the potential of a charged conductor of finite size we may connect the conductor with one electrode of the electrometer, while the other is connected to earth or to a body of constant potential. The electrometer reading will give the potential of the conductor after the division of its electricity between it and the part of the electrometer with which it is put in contact. If K denote the capacity of the conductor, and K' that of this part of the electrometer, and if V, V' denote the potentials of these bodies before making contact, then their common potential after making contact will be

$$\overline{V} = \frac{KV + K'V'}{K + K'}.$$

Hence the original potential of the conductor was

$$V = \overline{V} + \frac{K'}{K}(\overline{V} - V').$$

If the conductor is not large compared with the electrometer, K' will be comparable with K, and unless we can ascertain the values of K and K' the second term of the expression will have a doubtful value. But if we can make the potential of the electrode of the electrometer very nearly equal to that of the body before making contact, then the uncertainty of the values of K and K' will be of little consequence.

If we know the value of the potential of the body approximately, we may charge the electrode by means of a 'replenisher' or otherwise to this approximate potential, and the next experiment will give a closer approximation. In this way we may measure the potential of a conductor whose capacity is small compared with that of the electrometer.

To Measure the Potential at any Point in the Air.

207*.] *First Method.* Place a sphere, whose radius is small compared with the distance of electrified conductors, with its centre at the given point. Connect it by means of a fine wire with the earth, then insulate it, and carry it to an electrometer and ascertain the total charge on the sphere.

Then, if V be the potential at the given point, and a the radius of the sphere, the charge of the sphere will be $-Va = Q$, and if V' be the potential of the sphere as measured by an electrometer when placed in a room whose walls are connected with the earth, then

$$Q = V'a,$$

whence

$$V + V' = 0,$$

or the potential of the air at the point where the centre of the sphere was placed is equal but of opposite sign to the potential of the sphere after being connected to earth, then insulated, and brought into a room.

This method has been employed by M. Delmann of Creuznach in measuring the potential at a certain height above the earth's surface.*

Second Method. We have supposed the sphere placed at the given point and first connected to earth, and then insulated, and carried into a space surrounded with conducting matter at potential zero.

Now let us suppose a fine insulated wire carried from the electrode of the electrometer to the place where the potential is to be measured. Let the sphere be first discharged completely. This may be done by putting it into the inside of a vessel of the same metal which nearly surrounds it and making it touch the vessel. Now let the sphere thus discharged be carried to the end of the wire and made to touch it. Since the sphere is not electrified it will be at the potential of the air at the place. If the electrode wire is at the same potential it will not be affected by the contact, but if the electrode is at a different potential it will by contact with the sphere be made nearer to that of the air than it was before. By a succession of such operations, the sphere being alternately discharged and made to touch the electrode, the potential of the electrode of the electrometer will continually approach that of the air at the given point.

* [Compare Art. 50.]

208*.] To measure the potential of a conductor without touching it, we may measure the potential of the air at any point in the neighbourhood of the conductor, and calculate that of the conductor from the result. If there be a hollow nearly surrounded by the conductor, then the potential at any point of the air in this hollow will be very nearly that of the conductor.

In this way it has been ascertained by Sir W. Thomson that if two hollow conductors, one of copper and the other of zinc, are in metallic contact, then the potential of the air in the hollow surrounded by zinc is positive with reference to that of the air in the hollow surrounded by copper.

Third Method. If by any means we can cause a succession of small bodies to detach themselves from the end of the electrode, the potential of the electrode will approximate to that of the surrounding air. This may be done by causing shot, filings, sand, or water to drop out of a funnel or pipe connected with the electrode. The point at which the potential is measured is that at which the stream ceases to be continuous and breaks into separate parts or drops.

CHAPTER XII.

THE MEASUREMENT OF ELECTRIC RESISTANCE.

209*.] In the present state of electrical science, the determination of the electric resistance of a conductor may be considered as the cardinal operation in electricity, in the same sense that the determination of weight is the cardinal operation in chemistry.

The reason of this is that the determination in absolute measure of other electrical magnitudes, such as quantities of electricity, electromotive forces, currents, &c., requires in each case a complicated series of operations, involving generally observations of time, measurements of distances, and determinations of moments of inertia, and these operations, or at least some of them, must be repeated for every new determination, because it is impossible to preserve a unit of electricity, or of electromotive force, or of current, in an unchangeable state, so as to be available for direct comparison.

But when the electric resistance of a properly shaped conductor of a properly chosen material has been once determined, it is found that it always remains the same for the same temperature*, so that the conductor may be used as a standard of resistance, with which that of other conductors can be compared, and the comparison of two resistances is an operation which admits of extreme accuracy.

When the unit of electrical resistance has been fixed on, material copies of this unit, in the form of 'Resistance Coils,' are prepared for the use of electricians, so that in every part of the world electrical resistances may be expressed in terms of the same unit. These unit resistance coils are at present the only examples of material electric standards which can be preserved, copied, and used for the purpose of measurement. Measures of electrical capacity, which are also of great importance, are still defective, on account of the disturbing influence of electric absorption.

210*.] The unit of resistance may be an entirely arbitrary one, as in the case of Jacobi's Etalon, which was a certain copper

* [Recent observations have shewn that it is far from easy to find a material satisfying this condition.]

wire of 22·4932 grammes weight, 7·61975 metres length, and 0·667 millimetres diameter. Copies of this have been made by Leyser of Leipsig, and are to be found in different places.

According to another method the unit may be defined as the resistance of a portion of a definite substance of definite dimensions. Thus, Siemens' unit is defined as the resistance of a column of mercury of one metre long, and one square millimetre section, at the temperature 0°C.

211*.] Finally, the unit may be defined with reference to the electrostatic or the electromagnetic system of units. In practice the electromagnetic system is used in all telegraphic operations, and therefore the only systematic units actually in use are those of this system.

In the electromagnetic system a resistance is a quantity homogeneous with a velocity, and may therefore be expressed as a velocity.

212*.] The first actual measurements on this system were made by Weber, who employed as his unit one millimetre per second. Sir W. Thomson afterwards used one foot per second as a unit, but a large number of electricians have now agreed to use the unit of the British Association, which professes to represent a resistance which, expressed as a velocity, is ten millions of metres per second. The magnitude of this unit is more convenient than that of Weber's unit, which is too small. It is sometimes referred to as the B.A. unit, but in order to connect it with the name of the discoverer of the laws of resistance, it is called the Ohm.

213*.] To recollect its value in absolute measure it is useful to know that ten millions of metres is professedly the distance from the pole to the equator, measured along the meridian of Paris. A body, therefore, which in one second travels along a meridian from the pole to the equator would have a velocity which, on the electromagnetic system, is professedly represented by an Ohm.

I say professedly, because, if more accurate researches should prove that the Ohm, as constructed from the British Association's material standards, is not really represented by this velocity, electricians would not alter their standards, but would apply a correction. In the same way the metre is professedly one ten-millionth of a certain quadrantal arc, but though this is found not to be exactly true, the length of the metre has not been altered, but the dimensions of the earth are expressed by a less simple number.

According to the system of the British Association, the absolute value of the unit is *originally chosen* so as to represent as nearly as possible a quantity derived from the electromagnetic absolute system.

214*.] When a material unit representing this abstract quantity has been made, other standards are constructed by copying this unit, a process capable of extreme accuracy—of much greater accuracy than, for instance, the copying of foot-rules from a standard foot.

These copies, made of the most permanent materials, are distributed over all parts of the world, so that it is not likely that any difficulty will be found in obtaining copies of them if the original standards should be lost.

But such units as that of Siemens can without very great labour be reconstructed with considerable accuracy, so that as the relation of the Ohm to Siemens unit is known, the Ohm can be reproduced even without having a standard to copy, though the labour is much greater and the accuracy much less than by the method of copying.

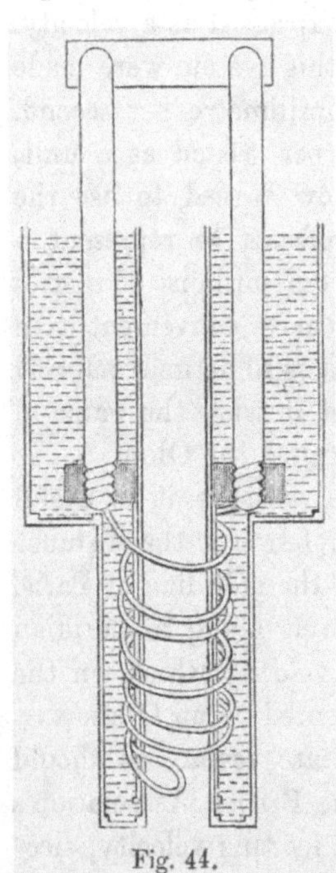

Fig. 44.

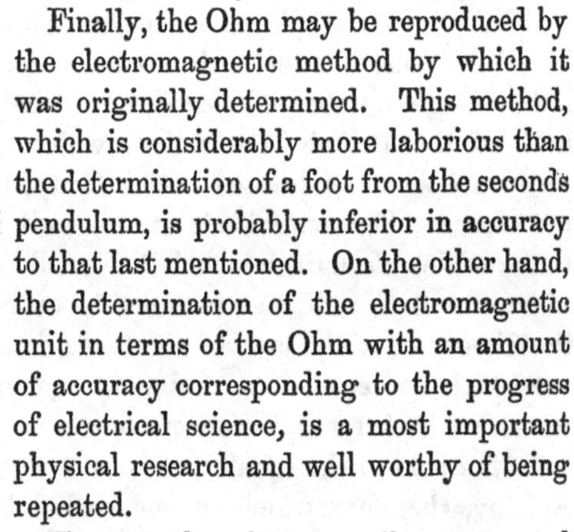

Finally, the Ohm may be reproduced by the electromagnetic method by which it was originally determined. This method, which is considerably more laborious than the determination of a foot from the seconds pendulum, is probably inferior in accuracy to that last mentioned. On the other hand, the determination of the electromagnetic unit in terms of the Ohm with an amount of accuracy corresponding to the progress of electrical science, is a most important physical research and well worthy of being repeated.

The actual resistance coils constructed to represent the Ohm were made of an alloy of two parts of silver and one of platinum in the form of wires from ·5 millimetres to ·8 millimetres diameter, and from one to two metres in length. These wires were soldered to stout copper electrodes. The wire itself was covered with two layers of silk, imbedded in solid paraffin, and enclosed in a thin brass case, so that it can be easily brought to a temperature at which its resistance is accurately one Ohm. This temperature is marked on the insulating support of the coil. (See Fig. 44.)

On the Forms of Resistance Coils.

215*.] A Resistance Coil is a conductor capable of being easily placed in the voltaic circuit, so as to introduce into the circuit a known resistance.

The electrodes or ends of the coil must be such that no appreciable error may arise from the mode of making the connexions. For resistances of considerable magnitude it is sufficient that the electrodes should be made of stout copper wire or rod well amalgamated with mercury at the ends, and that the ends should be made to press on flat amalgamated copper surfaces placed in mercury cups.

For very great resistances it is sufficient that the electrodes should be thick pieces of brass, and that the connexions should be made by inserting a wedge of brass or copper into the interval between them. This method is found very convenient.

The resistance coil itself consists of a wire well covered with silk, the ends of which are soldered permanently to the electrodes.

The coil must be so arranged that its temperature may be easily observed. For this purpose the wire is coiled on a tube and covered with another tube, so that it may be placed in a vessel of water, and that the water may have access to the inside and the outside of the coil.

To avoid the electromagnetic effects of the current in the coil the wire is first doubled back on itself and then coiled on the tube, so that at every part of the coil there are equal and opposite currents in the adjacent parts of the wire.

When it is desired to keep two coils at the same temperature the wires are sometimes placed side by side and coiled up together. This method is especially useful when it is more important to secure equality of resistance than to know the absolute value of the resistance, as in the case of the equal arms of Wheatstone's Bridge (Art. 221).

When measurements of resistance were first attempted, a resistance coil, consisting of an uncovered wire coiled in a spiral groove round a cylinder of insulating material, was much used. It was called a Rheostat. The accuracy with which it was found possible to compare resistances was soon found to be inconsistent with the use of any instrument in which the contacts are not more perfect than can be obtained in the rheostat. The rheostat, however, is still used for adjusting the resistance where accurate measurement is not required.

Resistance coils are generally made of those metals whose resistance is greatest and which vary least with temperature. German silver fulfils these conditions very well, but some specimens are found to change their properties during the lapse of years. Hence for standard coils, several pure metals, and also an alloy of platinum and silver, have been employed, and the relative resistance of these during several years has been found constant up to the limits of modern accuracy*.

216*.] For very great resistances, such as several millions of Ohms, the wire must be either very long or very thin, and the construction of the coil is expensive and difficult. Hence tellurium and selenium have been proposed as materials for constructing standards of great resistance. A very ingenious and easy method of construction has been lately proposed by Phillips†. On a piece of ebonite or ground glass a fine pencil-line is drawn. The ends of this filament of plumbago are connected to metallic electrodes, and the whole is then covered with insulating varnish. If it should be found that the resistance of such a pencil-line remains constant, this will be the best method of obtaining a resistance of several millions of Ohms.

217*.] There are various arrangements by which resistance coils may be easily introduced into a circuit.

For instance, a series of coils of which the resistances are 1, 2, 4, 8, 16, &c., arranged according to the powers of 2, may be placed in a box in series.

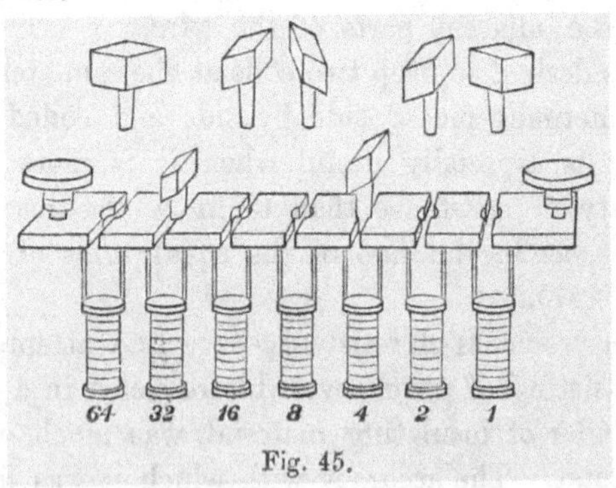

Fig. 45.

The electrodes consist of stout brass plates, so arranged on the

* [More recent experiments indicate a small change in resistance in course of time.]

† *Phil. Mag.*, July, 1870.

outside of the box that by inserting a brass plug or wedge between two of them as a shunt, the resistance of the corresponding coil may be put out of the circuit. This arrangement was introduced by Siemens.

Each interval between the electrodes is marked with the resistance of the corresponding coil, so that if we wish to make the resistance box equal to 107 we express 107 in the binary scale as 64+32+8+2+1 or 1101011. We then take the plugs out of the holes corresponding to 64, 32, 8, 2 and 1, and leave the plugs in 16 and 4.

This method, founded on the binary scale, is that in which the smallest number of separate coils is needed, and it is also that which can be most readily tested. For if we have another coil equal to 1 we can test the equality of 1 and $1'$, then that of $1+1'$ and 2, then that of $1+1'+2$ and 4, and so on.

The only disadvantage of the arrangement is that it requires a familiarity with the binary scale of notation, which is not generally possessed by those accustomed to express every number in the decimal scale.

218*.] A box of resistance coils may be arranged in a different way for the purpose of measuring conductivities instead of resistances.

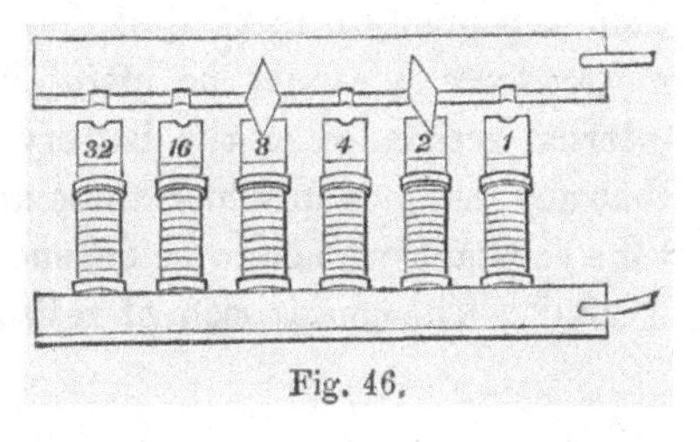

Fig. 46.

The coils are placed so that one end of each is connected with a long thick piece of metal which forms one electrode of the box, and the other end is connected with a stout piece of brass plate as in the former case.

The other electrode of the box is a long brass plate, such that by inserting brass plugs between it and the electrodes of the coils it may be connected to the first electrode through any given set of coils. The conductivity of the box is then the sum of the conductivities of the coils.

In the figure, in which the resistances of the coils are 1, 2, 4, &c., and the plugs are inserted at 2 and 8, the conductivity of the box is $\frac{1}{2}+\frac{1}{8}=\frac{5}{8}$, and the resistance of the box is therefore $\frac{8}{5}$ or 1·6.

This method of combining resistance coils for the measurement of fractional resistances was introduced by Sir W. Thomson under the name of the method of multiple arcs. See Art. 158.

On the Comparison of Resistances.

219*.] If E is the electromotive force of a battery, and R the resistance of the battery and its connexions, including the galvanometer used in measuring the current, and if the strength of the current is I when the battery connexions are closed, and I_1, I_2 when additional resistances r_1, r_2 are introduced into the circuit, then, by Ohm's Law,

$$E = IR = I_1(R+r_1) = I_2(R+r_2).$$

Eliminating E, the electromotive force of the battery, and R the resistance of the battery and its connexions, we get Ohm's formula

$$\frac{r_1}{r_2} = \frac{(I-I_1)I_2}{(I-I_2)I_1}.$$

This method requires a measurement of the ratios of I, I_1 and I_2, and this implies a galvanometer graduated for absolute measurements.

If the resistances r_1 and r_2 are equal, then I_1 and I_2 are equal, and we can test the equality of currents by a galvanometer which is not capable of determining their ratios.

But this is rather to be taken as an example of a faulty method than as a practical method of determining resistance. The electromotive force E cannot be maintained rigorously constant, and the internal resistance of the battery is also exceedingly variable, so that any methods in which these are assumed to be even for a short time constant are not to be depended on.

220*.] The comparison of resistances can be made with extreme

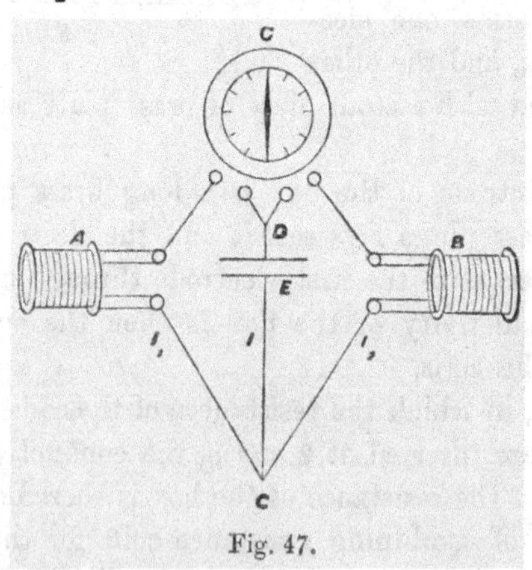

Fig. 47.

accuracy by either of two methods, in which the result is independent of variations of R and E.

The first of these methods depends on the use of the differential galvanometer, an instrument in which there are two coils, the currents in which are independent of each other, so that when the currents are made to flow in opposite directions they act in opposite directions on the needle, and when the ratio of these currents is that of m to n they have no resultant effect on the galvanometer needle.

Let I_1, I_2 be the currents through the two coils of the galvanometer, then the deflexion of the needle may be written

$$\delta = mI_1 - nI_2.$$

Now let the battery current I be divided between the coils of the galvanometer, and let resistances A and B be introduced into the first and second coils respectively. Let the remainder of the resistance of their coils and their connexions be α and β respectively, and let the resistance of the battery and its connexions between C and D be r, and its electromotive force E.

Then we find, by Ohm's Law, for the difference of potentials between C and D,

$$C - D = I_1(A+\alpha) = I_2(B+\beta) = E - Ir,$$

and since

$$I_1 + I_2 = I,$$

$$I_1 = E\frac{B+\beta}{D}, \qquad I_2 = E\frac{A+\alpha}{D}, \qquad I = E\frac{A+\alpha+B+\beta}{D},$$

where

$$D = (A+\alpha)(B+\beta) + r(A+\alpha+B+\beta).$$

The deflexion of the galvanometer needle is therefore

$$\delta = \frac{E}{D}\{m(B+\beta) - n(A+\alpha)\},$$

and if there is no observable deflexion, then we know that the quantity enclosed in brackets cannot differ from zero by more than a certain small quantity, depending on the power of the battery, the suitableness of the arrangement, the delicacy of the galvanometer, and the accuracy of the observer.

Suppose that B has been adjusted so that there is no apparent deflexion.

Now let another conductor A' be substituted for A, and let A' be adjusted till there is no apparent deflexion. Then evidently to a first approximation $A' = A$.

To ascertain the degree of accuracy of this estimate, let the altered quantities in the second observation be accented, then

$$m(B+\beta)-n(A+\alpha)=\frac{D}{E}\delta,$$

$$m(B+\beta)-n(A'+\alpha)=\frac{D'}{E'}\delta'.$$

Hence $$n(A'-A)=\frac{D}{E}\delta-\frac{D'}{E'}\delta'.$$

If δ and δ', instead of being both apparently zero, had been only observed to be equal, then, unless we also could assert that $E=E'$, the right-hand side of the equation might not be zero. In fact, the method would be a mere modification of that already described.

The merit of the method consists in the fact that the thing observed is the absence of any deflexion, or in other words, the method is a Null method, one in which the non-existence of a force is asserted from an observation in which the force, if it had been different from zero by more than a certain small amount, would have produced an observable effect.

Null methods are of great value where they can be employed, but they can only be employed where we can cause two equal and opposite quantities of the same kind to enter into the experiment together.

In the case before us both δ and δ' are quantities too small to be observed, and therefore any change in the value of E will not affect the accuracy of the result.

The actual degree of accuracy of this method might be ascertained by taking a number of observations in each of which A' is separately adjusted, and comparing the result of each observation with the mean of the whole series.

But by putting A out of adjustment by a known quantity, as, for instance, by inserting at A or at B an additional resistance equal to a hundredth part of A or of B, and then observing the resulting deviation of the galvanometer needle, we can estimate the number of degrees corresponding to an error of one per cent. To find the actual degree of precision we must estimate the smallest deflexion which could not escape observation, and compare it with the deflexion due to an error of one per cent.

*If the comparison is to be made between A and B, and if the positions of A and B are exchanged, then the second equation becomes

* This investigation is taken from Weber's treatise on Galvanometry. *Göttingen Transactions*, x. p. 65.

$$m(A+\beta)-n(B+a)=\frac{D'}{E'}\delta',$$

whence $$(m+n)(B-A)=\frac{D}{E}\delta-\frac{D'}{E'}\delta'.$$

If m and n, A and B, a and β are approximately equal, then

$$B-A=\frac{1}{2nE}(A+a)(A+a+2r)(\delta-\delta').$$

Here $\delta-\delta'$ may be taken to be the smallest observable deflexion of the galvanometer.

If the galvanometer wire be made longer and thinner, retaining the same total mass, then n will vary as the length of the wire and a as the square of the length. Hence there will be a minimum value of $\frac{(A+a)(A+a+2r)}{n}$ when

$$a=\tfrac{1}{3}(A+r)\left\{2\sqrt{1-\frac{3}{4}\frac{r^2}{(A+r)^2}}-1\right\}.$$

If we suppose r, the battery resistance, small compared with A, this gives $$a=\tfrac{1}{3}A;$$

or, *the resistance of each coil of the galvanometer should be one-third of the resistance to be measured.*

We then find $$B-A=\frac{8}{9}\frac{A^2}{nE}(\delta-\delta').$$

If we allow the current to flow through one only of the coils of the galvanometer, and if the deflexion thereby produced is Δ (supposing the deflexion strictly proportional to the deflecting force), then

$$\Delta=\frac{mE}{A+a+r}=\frac{3}{4}\frac{nE}{A}\text{ if } r=0 \text{ and } a=\frac{1}{3}A.$$

Hence $$\frac{B-A}{A}=\frac{2}{3}\frac{\delta-\delta'}{\Delta}.$$

In the differential galvanometer two currents are made to produce equal and opposite effects on the suspended needle. The force with which either current acts on the needle depends not only on the strength of the current, but on the position of the windings of the wire with respect to the needle. Hence, unless the coil is very carefully wound, the ratio of m to n may change when the position of the needle is changed, and therefore it is necessary to determine this ratio by proper methods during each course of experiments if any alteration of the position of the needle is suspected.

The other null method, in which Wheatstone's Bridge is used, requires only an ordinary galvanometer, and the observed zero deflexion of the needle is due, not to the opposing action of two currents, but to the non-existence of a current in the wire. Hence we have not merely a null deflexion, but a null current as the phenomenon observed, and no errors can arise from want of regularity or change of any kind in the coils of the galvanometer. The galvanometer is only required to be sensitive enough to detect the existence and direction of a current, without in any way determining its value or comparing its value with that of another current.

221*.] Wheatstone's Bridge consists essentially of six conductors connecting four points. An electromotive force E is made to act between two of the points by means of a voltaic battery introduced between B and C. The current between the other two points O and A is measured by a galvanometer.

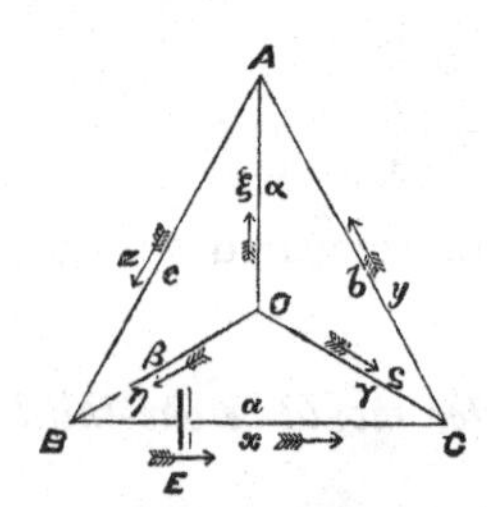

Fig. 48.

Under certain circumstances this current becomes zero. The conductors BC and OA are then said to be *conjugate* to each other, which implies a certain relation between the resistances of the other four conductors, and this relation is made use of in measuring resistances.

If the current in OA is zero, the potential at O must be equal to that at A. Now when we know the potentials at B and C we can determine those at O and A by the rule given at Art. 157, provided there is no current in OA,

$$O = \frac{B\gamma + C\beta}{\beta + \gamma}, \quad A = \frac{Bb + Cc}{b + c},$$

whence the condition is $$b\beta = c\gamma,$$

where b, c, β, γ are the resistances in CA, AB, BO and OC respectively.

To determine the degree of accuracy attainable by this method we must ascertain the strength of the current in OA when this condition is not fulfilled exactly.

Let A, B, C and O be the four points. Let the currents along BC, CA and AB be x, y and z, and the resistances of these conductors a, b and c. Let the currents along OA, OB and OC be ξ, η, ζ and the resistances α, β and γ. Let an electromotive force E act along BC. Required the current ξ along OA.

Let the potentials at the points A, B, C and O be denoted by the symbols A, B, C and O. The equations of conduction are

$$\begin{aligned} ax &= B-C+E, & \alpha\xi &= O-A, \\ by &= C-A & \beta\eta &= O-B, \\ cz &= A-B & \gamma\zeta &= O-C; \end{aligned}$$

with the equations of continuity

$$\begin{aligned} \xi+y-z &= 0, \\ \eta+z-x &= 0, \\ \zeta+x-y &= 0. \end{aligned}$$

By considering the system as made up of three circuits OBC, OCA and OAB in which the currents are x, y, z respectively, and applying Kirchhoff's rule [Art. 158] to each cycle, we eliminate the values of the potentials O, A, B, C, and the currents ξ, η, ζ and obtain the following equations for x, y and z,

$$\begin{aligned} (a+\beta+\gamma)x - \gamma y \quad - \beta z &= E, \\ -\gamma x + (b+\gamma+\alpha)y - \alpha z &= 0, \\ -\beta x - \alpha y + (c+\alpha+\beta)z &= 0. \end{aligned}$$

Hence, if we put

$$D = \begin{vmatrix} a+\beta+\gamma & -\gamma & -\beta \\ -\gamma & b+\gamma+\alpha & -\alpha \\ -\beta & -\alpha & c+\alpha+\beta \end{vmatrix},$$

we find
$$\xi = \frac{E}{D}(b\beta - c\gamma),$$

and
$$x = \frac{E}{D}\{(b+\gamma)(c+\beta) + \alpha(b+c+\beta+\gamma)\}.$$

222*.] The value of D may be expressed in the symmetrical form,

$$D = abc + bc(\beta+\gamma) + ca(\gamma+\alpha) + ab(\alpha+\beta) + (a+b+c)(\beta\gamma+\gamma\alpha+\alpha\beta)$$

or, since we suppose the battery in the conductor a and the galvanometer in α, we may put B the battery resistance for a and G the galvanometer resistance for α. We then find

$$D = BG(b+c+\beta+\gamma) + B(b+\gamma)(c+\beta) \\ + G(b+c)(\beta+\gamma) + bc(\beta+\gamma) + \beta\gamma(b+c).$$

If the electromotive force E were made to act along OA, the resistance of OA being still α, and if the galvanometer were placed in BC, the resistance of BC being still a, then the value of D would remain the same, and the current in BC due to the electromotive force E acting along OA would be equal to the current in OA due to the electromotive force E acting in BC.

But if we simply disconnect the battery and the galvanometer, and without altering their respective resistances connect the battery to O and A and the galvanometer to B and C, then in the value of D we must exchange the values of B and G. If D' be the value of D after this exchange, we find

$$D'-D = (G-B)\{(b+c)(\beta+\gamma)-(b+\gamma)(\beta+c)\},$$
$$= (B-G)\{(b-\beta)(c-\gamma)\}.$$

Let us suppose that the resistance of the galvanometer is greater than that of the battery.

Let us also suppose that in its original position the galvanometer connects the junction of the two conductors of least resistance β, γ with the junction of the two conductors of greatest resistance b, c, or, in other words, we shall suppose that if the quantities b, c, γ, β are arranged in order of magnitude, b and c stand together, and γ and β stand together. Hence the quantities $b-\beta$ and $c-\gamma$ are of the same sign, so that their product is positive, and therefore $D'-D$ is of the same sign as $B-G$.

If therefore the galvanometer is made to connect the junction of the two greatest resistances with that of the two least, and if the galvanometer resistance is greater than that of the battery, then the value of D will be less, and the value of the deflexion of the galvanometer greater, than if the connexions are exchanged.

The rule therefore for obtaining the greatest galvanometer deflexion in a given system is as follows:

Of the two resistances, that of the battery and that of the galvanometer, connect the greater resistance so as to join the two greatest to the two least of the four other resistances.

223*.] We shall suppose that we have to determine the ratio of the resistances of the conductors AB and AC, and that this is to be done by finding a point O on the conductor BOC, such that when the points A and O are connected by a wire, in the course of which a galvanometer is inserted, no sensible deflexion of the galvanometer needle occurs when the battery is made to act between B and C.

The conductor BOC may be supposed to be a wire of uniform resistance divided into equal parts, so that the ratio of the resistances of BO and OC may be read off at once.

Instead of the whole conductor being a uniform wire, we may make the part near O of such a wire, and the parts on each side may be coils of any form, the resistance of which is accurately known.

We shall now use a different notation instead of the symmetrical notation with which we commenced.

Let the whole resistance of BAC be R.

Let $c = mR$ and $b = (1-m)R$.

Let the whole resistance of BOC be S.

Let $\beta = nS$ and $\gamma = (1-n)S$.

The value of n is read off directly, and that of m is deduced from it when there is no sensible deviation of the galvanometer.

Let the resistance of the battery and its connexions be B, and that of the galvanometer and its connections G.

We find as before

$$D = G\{BR+BS+RS\} + m(1-m)R^2(B+S) + n(1-n)S^2(B+R) + (m+n-2mn)BRS,$$

and if ξ is the current in the galvanometer wire

$$\xi = \frac{ERS}{D}(n-m).$$

In order to obtain the most accurate results we must make the deviation of the needle as great as possible compared with the value of $(n-m)$. This may be done by properly choosing the dimensions of the galvanometer and the standard resistance wire.

It may be shewn that when the form of a galvanometer wire is changed while its mass remains constant, the deviation of the needle for unit current is proportional to the length, but the resistance increases as the square of the length. Hence the maximum deflexion is shewn to occur when the resistance of the galvanometer wire is equal to the constant resistance of the rest of the circuit.

In the present case, if δ is the deviation,

$$\delta = C\sqrt{G}\,\xi,$$

where C is some constant, and G is the galvanometer resistance which varies as the square of the length of the wire. Hence we find that in the value of D, when δ is a maximum, the part involving G must be made equal to the rest of the expression.

If we also put $m = n$, as is the case if we have made a correct observation, we find the best value of G to be

$$G = n(1-n)(R+S).$$

This result is easily obtained by considering the resistance from A to O through the system, remembering that BC, being conjugate to AO, has no effect on this resistance.

In the same way we should find that if the total area of the acting surfaces of the battery is given, the most advantageous arrangement of the battery is when

$$B = \frac{RS}{R+S}.$$

Finally, we shall determine the value of S such that a given change in the value of n may produce the greatest galvanometer deflexion. By differentiating the expression for ξ we find

$$S^2 = \frac{BR}{B+R}\left(R + \frac{G}{n(1-n)}\right).$$

If we have a great many determinations of resistance to make in which the actual resistance has nearly the same value, then it may be worth while to prepare a galvanometer and a battery for this purpose. In this case we find that the best arrangement is

$$S = R, \qquad B = \tfrac{1}{2}R, \qquad G = 2n(1-n)R,$$

and if $n = \frac{1}{2}$, $G = \frac{1}{2}R$.

On the Use of Wheatstone's Bridge.

224*.] We have already explained the general theory of Wheatstone's Bridge, we shall now consider some of its applications.

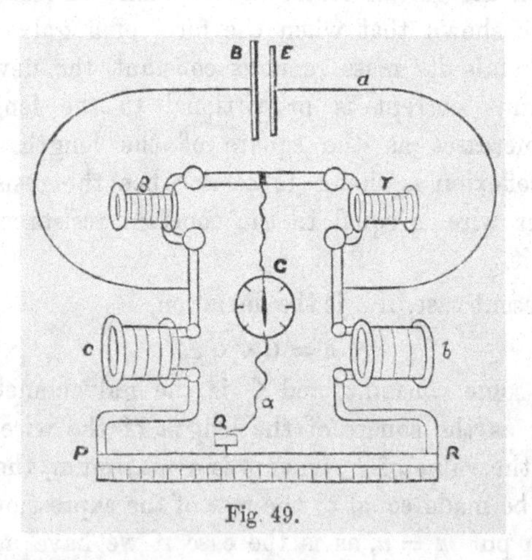

Fig. 49.

The comparison which can be effected with the greatest exactness is that of two equal resistances.

Let us suppose that β is a standard resistance coil, and that we wish to adjust γ to be equal in resistance to β.

Two other coils, b and c, are prepared which are equal or nearly equal to each other, and the four coils are placed with their electrodes in mercury cups so that the current of the battery is divided between two branches, one consisting of β and γ and the other of b and c. The coils b and c are connected by a wire PR, as uniform in its resistance as possible, and furnished with a scale of equal parts.

The galvanometer wire connects the junction of β and γ with a point Q of the wire PR, and the point of contact at Q is made to vary till on closing first the battery circuit and then the galvanometer circuit, no deflexion of the galvanometer needle is observed.

The coils β and γ are then made to change places, and a new position is found for Q. If this new position is the same as the old one, then we know that the exchange of β and γ has produced no change in the proportions of the resistances, and therefore γ is rightly adjusted. If Q has to be moved, the direction and amount of the change will indicate the nature and amount of the alteration of the length of the wire of γ, which will make its resistance equal to that of β.

If the resistances of the coils b and c, each including part of the wire PR up to its zero reading, are equal to that of b and c divisions of the wire respectively, then, if x is the scale reading of Q in the first case, and y that in the second,

$$\frac{c+x}{b-x} = \frac{\beta}{\gamma}, \qquad \frac{c+y}{b-y} = \frac{\gamma}{\beta},$$

whence

$$\frac{\gamma^2}{\beta^2} = 1 + \frac{(b+c)(y-x)}{(c+x)(b-y)}.$$

Since $b-y$ is nearly equal to $c+x$, and both are great with respect to x or y, we may write this

$$\frac{\gamma^2}{\beta^2} = 1 + 4\frac{y-x}{b+c},$$

and

$$\gamma = \beta\left(1 + 2\frac{y-x}{b+c}\right).$$

When γ is adjusted as well as we can, we substitute for b and c other coils of (say) ten times greater resistance.

The remaining difference between β and γ will now produce a ten times greater difference in the position of Q than with the original coils b and c, and in this way we can continually increase the accuracy of the comparison.

The adjustment by means of the wire with sliding contact piece is more quickly made than by means of a resistance box, and it is capable of continuous variation.

The battery must never be introduced instead of the galvanometer into the wire with a sliding contact, for the passage of a powerful current at the point of contact would injure the surface of the wire. Hence this arrangement is adapted for the case in which the resistance of the galvanometer is greater than that of the battery.

When γ, the resistance to be measured, a, the resistance of the battery, and α, the resistance of the galvanometer, are given, the best values of the other resistances have been shewn by Mr. Oliver Heaviside (*Phil. Mag.*, Feb. 1873) to be

$$c = \sqrt{a\alpha}, \quad b = \sqrt{a\gamma\frac{\alpha+\gamma}{a+\gamma}}, \quad \beta = \sqrt{\alpha\gamma\frac{a+\gamma}{\alpha+\gamma}}.$$

Thomson's Method for the Determination of the Resistance of the Galvanometer.*

225*.] An arrangement similar to Wheatstone's Bridge has been employed with advantage by Sir W. Thomson in determining the resistance of the galvanometer when in actual use. It was suggested to Sir W. Thomson by Mance's Method. See Art. 226.

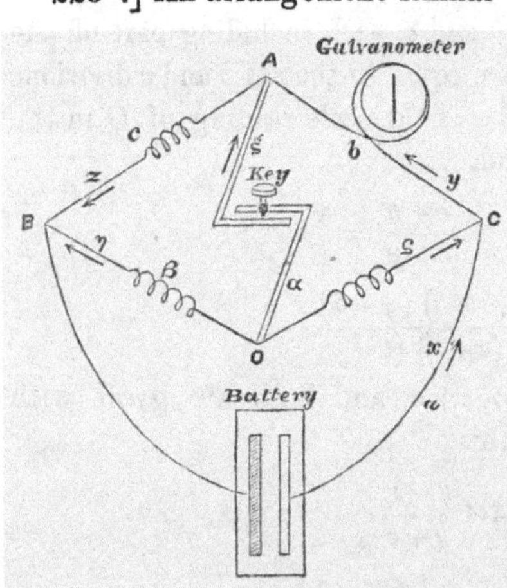

Fig. 50.

Let the battery be placed, as before, between B and C in the figure of Article 221, but let the galvanometer be placed in CA instead of in OA. If $b\beta - c\gamma$ is zero, then the conductor OA is conjugate to BC, and, as there is no current produced in OA by the battery in BC, the strength of the current in any other conductor is independent of the resistance

* *Proc. R. S.*, Jan. 19, 1871.

in OA. Hence, if the galvanometer is placed in CA its deflexion will remain the same whether the resistance of OA is small or great. We therefore observe whether the deflexion of the galvanometer remains the same when O and A are joined by a conductor of small resistance, as when this connexion is broken, and if, by properly adjusting the resistances of the conductors, we obtain this result, we know that the resistance of the galvanometer is

$$b = \frac{c\gamma}{\beta}$$

where c, γ, and β are resistance coils of known resistance.

It will be observed that though this is not a null method, in the sense of there being no current in the galvanometer, it is so in the sense of the fact observed being the negative one, that the deflexion of the galvanometer is not changed when a certain contact is made. An observation of this kind is of greater value than an observation of the equality of two different deflexions of the same galvanometer, for in the latter case there is time for alteration in the strength of the battery or the sensitiveness of the galvanometer, whereas when the deflexion remains constant, in spite of certain changes which we can repeat at pleasure, we are sure that the current is quite independent of these changes.

The determination of the resistance of the coil of a galvanometer can easily be effected in the ordinary way of using Wheatstone's Bridge by placing another galvanometer in OA. By the method now described the galvanometer itself is employed to measure its own resistance.

Mance's Method of determining the Resistance of the Battery.*

226*.] The measurement of the resistance of a battery when in action is of a much higher order of difficulty, since the resistance of the battery is found to change considerably for some time after the strength of the current through it is changed. In many of the methods commonly used to measure the resistance of a battery such alterations of the strength of the current through it occur in the course of the operations, and therefore the results are rendered doubtful.

In Mance's method, which is free from this objection, the battery is placed in BC and the galvanometer in CA. The connexion between O and B is then alternately made and broken.

* *Proc. R. S.*, Jan. 19, 1871.

If the deflexion of the galvanometer remains unaltered, we know that OB is conjugate to CA, whence $c\gamma = a\alpha$, and a, the resistance of the battery, is obtained in terms of known resistances c, γ, α.

When the condition $c\gamma = a\alpha$ is fulfilled, then the current through the galvanometer is

$$y = \frac{Ea}{ba + c(b + a + \gamma)},$$

and this is independent of the resistance β between O and B. To test the sensibility of the method let us suppose that the condition $c\gamma = a\alpha$ is nearly, but not accurately, fulfilled, and that y_0 is the current through the galvanometer when O and B are connected

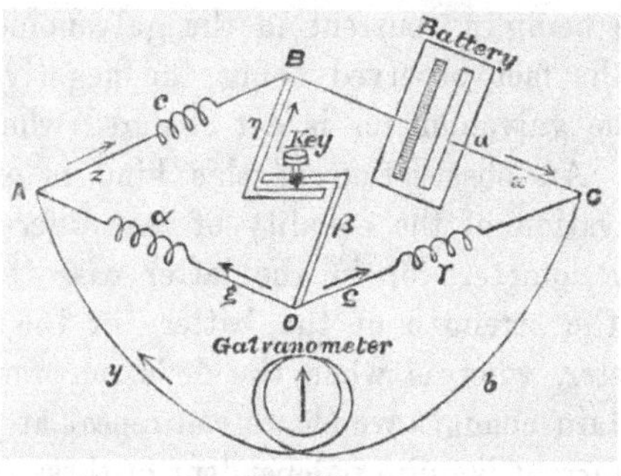

Fig. 51.

by a conductor of no sensible resistance, and y_1 the current when O and B are completely disconnected.

To find these values we must make β equal to 0 and to ∞ in the general formula for y, and compare the results.

In this way we find

$$\frac{y_0 - y_1}{y} = \frac{\alpha}{\gamma} \frac{c\gamma - a\alpha}{(c + a)(\alpha + \gamma)},$$

where y_0 and y_1 are supposed to be so nearly equal that we may, when their difference is not in question, put either of them equal to y, the value of the current when the adjustment is perfect.

The resistance, c, of the conductor AB should be equal to a, that of the battery, α and γ, should be equal and as small as possible, and b should be equal to $a + \gamma$.

Since a galvanometer is most sensitive when its deflexion is small, we should bring the needle nearly to zero by means of fixed magnets before making contact between O and B.

In this method of measuring the resistance of the battery, the current in the battery is not in any way interfered with during the operation, so that we may ascertain its resistance for any given strength of current, so as to determine how the strength of current affects the resistance.

If y is the current in the galvanometer, the actual current through the battery is x_0 with the key down and x_1 with the key up, where

$$x_0 = y\left(1 + \frac{b}{a+\gamma}\right), \qquad x_1 = y\left(1 + \frac{b}{\gamma} + \frac{ac}{\gamma(a+c)}\right),$$

the resistance of the battery is

$$a = \frac{c\gamma}{a},$$

and the electromotive force of the battery is

$$E = y\left(b + c + \frac{c}{a}(b+\gamma)\right).$$

The method of Art. 225 for finding the resistance of the galvanometer differs from this only in making and breaking contact between O and A instead of between O and B, and by exchanging a and β we obtain for this case

$$\frac{y_0 - y_1}{y} = \frac{\beta}{\gamma} \frac{c\gamma - b\beta}{(c+\beta)(\beta+\gamma)}.$$

On the Comparison of Electromotive Forces.

227*.] The following method of comparing the electromotive forces of voltaic and thermoelectric arrangements, when no current passes through them, requires only a set of resistance coils and a constant battery.

Let the electromotive force E of the battery be greater than that of either of the electromotors to be compared, then, if a sufficient

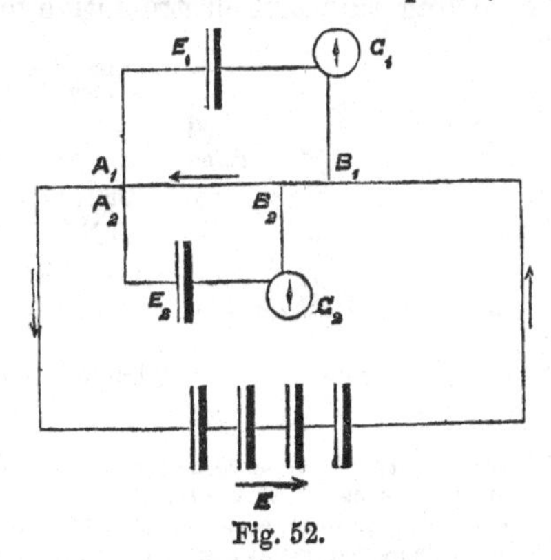

Fig. 52.

resistance, R_1, be interposed between the points A_1, B_1 of the primary circuit EB_1A_1E, the electromotive force from B_1 to A_1

may be made equal to that of the electromotor E_1. If the electrodes of this electromotor are now connected with the points A_1, B_1 no current will flow through the electromotor. By placing a galvanometer G_1 in the circuit of the electromotor E_1, and adjusting the resistance between A_1 and B_1, till the galvanometer G_1 indicates no current, we obtain the equation

$$E_1 = R_1 C,$$

where R_1 is the resistance between A_1 and B_1, and C is the strength of the current in the primary circuit.

In the same way, by taking a second electromotor E_2 and placing its electrodes at A_2 and B_2, so that no current is indicated by the galvanometer G_2,

$$E_2 = R_2 C,$$

where R_2 is the resistance between A_2 and B_2. If the observations of the galvanometers G_1 and G_2 are simultaneous, the value of C, the current in the primary circuit, is the same in both equations, and we find

$$E_1 : E_2 :: R_1 : R_2.$$

In this way the electromotive force of two electromotors may be compared.* The absolute electromotive force of an electromotor may be measured either electrostatically by means of the electrometer, or electromagnetically by means of an absolute galvanometer.

This method, in which, at the time of the comparison, there is no current through either of the electromotors, is a modification of Poggendorff's method, and is due to Mr. Latimer Clark, who has deduced the following values of electromotive forces :

			Concentrated solution of		Volts.
Daniell I.	Amalgamated Zinc	$H_2SO_4 +$ 4 aq.	Cu SO_4	Copper	=1.079
II.	„	H_2SO + 12 aq.	Cu SO_4	Copper	=0.978
III.	„	H_2SO_4 + 12 aq.	Cu2 (NO_3)	Copper	=1.00
Bunsen I.	„	„ „	H NO_3	Carbon	=1.964
II.	„	„ „	sp. g. 1. 38	Carbon	=1.888
Grove	„	$H_2SO_4 +$ 4 aq.	H NO_3	Platinum	=1.956

A Volt is an electromotive force equal to 100,000,000 *units of the centimetre-gramme-second system.*

* [Any number of batteries may be compared by the help of only one galvanometer if one pole of each battery is connected with the same electrode of the galvanometer the other poles being connected through separate keys to points A_1, A_2, &c. upon the wire and the keys being depressed one at a time but in rapid succession.]

CHAPTER XIII.

ON THE ELECTRIC RESISTANCE OF SUBSTANCES.

228*.] There are three classes in which we may place different substances in relation to the passage of electricity through them.

The first class contains all the metals and their alloys, some sulphurets, and other compounds containing metals, to which we must add carbon in the form of gas-coke, and selenium in the crystalline form.

In all these substances conduction takes place without any decomposition, or alteration of the chemical nature of the substance, either in its interior or where the current enters and leaves the body. In all of them the resistance increases as the temperature rises.

The second class consists of substances which are called electrolytes, because the current is associated with a decomposition of the substance into two components which appear at the electrodes. As a rule a substance is an electrolyte only when in the liquid form, though certain colloid substances, such as glass at 100°C, which are apparently solid, are electrolytes. It would appear from the experiments of Sir B. C. Brodie that certain gases are capable of electrolysis by a powerful electromotive force.

In all substances which conduct by electrolysis the resistance diminishes as the temperature rises.

The third class consists of substances the resistance of which is so great that it is only by the most refined methods that the passage of electricity through them can be detected. These are called Dielectrics. To this class belong a considerable number of solid bodies, many of which are electrolytes when melted, some liquids, such as turpentine, naphtha, melted paraffin, &c., and all gases and vapours. Carbon in the form of diamond, and selenium in the amorphous form, belong to this class.

The resistance of this class of bodies is enormous compared with that of the metals. It diminishes as the temperature rises. It is difficult, on account of the great resistance of these substances, to determine whether the feeble current which we can force through them is or is not associated with electrolysis.

On the Electric Resistance of Metals.

229*.] There is no part of electrical research in which more numerous or more accurate experiments have been made than in the determination of the resistance of metals. It is of the utmost importance in the electric telegraph that the metal of which the wires are made should have the smallest attainable resistance. Measurements of resistance must therefore be made before selecting the materials. When any fault occurs in the line, its position is at once ascertained by measurements of resistance, and these measurements, in which so many persons are now employed, require the use of resistance coils, made of metal the electrical properties of which have been carefully tested.

The electrical properties of metals and their alloys have been studied with great care by MM. Matthiessen, Vogt, and Hockin, and by MM. Siemens, who have done so much to introduce exact electrical measurements into practical work.

It appears from the researches of Dr. Matthiessen, that the effect of temperature on the resistance is nearly the same for a considerable number of the *pure* metals, the resistance at 100°C being to that at 0°C in the ratio of 1.414 to 1, or of 1 to .707. For pure iron the ratio is 1.645, and for pure thallium 1.458.

The resistance of metals has been observed by Dr. C. W. Siemens* through a much wider range of temperature, extending from the freezing point to 350°C, and in certain cases to 1000°C. He finds that the resistance increases as the temperature rises, but that the rate of increase diminishes as the temperature rises. The formula, which he finds to agree very closely both with the resistances observed at low temperatures by Dr. Matthiessen and with his own observations through a range of 1000°C, is

$$r = \alpha T^{\frac{1}{2}} + \beta T + \gamma,$$

where T is the absolute temperature reckoned from -273°C, and α, β, γ are constants. Thus, for

Platinum......$r = 0.039369 T^{\frac{1}{2}} + 0.00216407 T - 0.2413$,

Copper.........$r = 0.026577 T^{\frac{1}{2}} + 0.0031443 T - 0.22751$,

Iron...........$r = 0.072545 T^{\frac{1}{2}} + 0.0038133 T - 1.23971$.

* *Proc. R. S.*, April, 27, 1871.

From data of this kind the temperature of a furnace may be determined by means of an observation of the resistance of a platinum wire placed in the furnace.

Dr. Matthiessen found that when two metals are combined to form an alloy, the resistance of the alloy is in most cases greater than that calculated from the resistance of the component metals and their proportions. In the case of alloys of gold and silver, the resistance of the alloy is greater than that of either pure gold or pure silver, and, within certain limiting proportions of the constituents, it varies very little with a slight alteration of the proportions. For this reason Dr. Matthiessen recommended an alloy of two parts by weight of gold and one of silver as a material for reproducing the unit of resistance.

The effect of change of temperature on electric resistance is generally less in alloys than in pure metals.

Hence ordinary resistance coils are made of German silver, on account of its great resistance and its small variation with temperature.

An alloy of silver and platinum is also used for standard coils.

230*.] In the following table R is the resistance in Ohms of a column one metre long and one gramme weight at 0°C, and r is the resistance in centimetres per second of a cube of one centimetre, according to the experiments of Matthiessen*.

	Specific gravity		R	r	Percentage increment of resistance for 1°C at 20°C.
Silver	10·50	hard drawn	0·1689	1609	0·377
Copper.........	8·95	hard drawn	0·1469	1642	0·388
Gold...........	19·27	hard drawn	0·4150	2154	0·365
Lead	11·391	pressed	2·257	19847	0·387
Mercury	13·595	liquid	13·071	96146	0·072
Gold 2, Silver 1 ..	15·218	hard or annealed	1·668	10988	0·065
Selenium at 100°C		Crystalline form		6×10^{13}	1·00

It appears from the researches of Matthiessen and Hockin that the resistance of a uniform column of mercury of one metre in length, and weighing one gramme at 0°C, is 13·071 Ohms, whence it follows that if the specific gravity of mercury is 13·595, the resistance of a column of one metre in length and one square millimetre in section is 0.96146 Ohms.

* *Phil. Mag.*, May, 1865.

On the Electric Resistance of Electrolytes.

231*.] The measurement of the electric resistance of electrolytes is rendered difficult on account of the polarization of the electrodes, which causes the observed difference of potentials of the metallic electrodes to be greater than the electromotive force which actually produces the current.

This difficulty can be overcome in various ways. In certain cases we can get rid of polarization by using electrodes of proper material, as, for instance, zinc electrodes in a solution of sulphate of zinc. By making the surface of the electrodes very large compared with the section of the part of the electrolyte whose resistance is to be measured, and by using only currents of short duration in opposite directions alternately, we can make the measurements before any considerable intensity of polarization has been excited by the passage of the current.

Finally, by making two different experiments, in one of which the path of the current through the electrolyte is much longer than in the other, and so adjusting the electromotive force that the actual current, and the time during which it flows, are nearly the same in each case, we can eliminate the effect of polarization altogether.

232*.] In the experiments of Dr. Paalzow * the electrodes were in the form of large disks placed in separate flat vessels filled with the electrolyte, and the connexion was made by means of a long siphon filled with the electrolyte and dipping into both vessels. Two such siphons of different lengths were used.

The observed resistances of the electrolyte in these siphons being R_1 and R_2, the siphons were next filled with mercury, and their resistances when filled with mercury were found to be R_1' and R_2'.

The ratio of the resistance of the electrolyte to that of a mass of mercury at 0°C of the same form was then found from the formula

$$\rho = \frac{R_1 - R_2}{R_1' - R_2'}$$

To deduce from the values of ρ the resistance of a centimetre in

* Berlin *Monatsbericht*, July, 1868.

length having a section of a square centimetre, we must multiply them by the value of r for mercury at 0°C. See Art. 230.

The results given by Paalzow are as follow :—

Mixtures of Sulphuric Acid and Water.

	Temp.	Resistance compared with mercury.
H_2SO_4	15°C	96950
$H_2SO_4 + 14\,H^2O$	19°C	14157
$H_2SO_4 + 13\,H^2O$	22°C	13310
$H_2SO_4 + 499\,H^2O$	22°C	184773

Sulphate of Zinc and Water.

$ZnSO_4 + 23\,H^2O$	23°C	194400
$ZnSO_4 + 24\,H^2O$	23°C	191000
$ZnSO_4 + 105\,H^2O$	23°C	354000

Sulphate of Copper and Water.

$CuSO_4 + 45\,H^2O$	22°C	202410
$CuSO_4 + 105\,H^2O$	22°C	339341

Sulphate of Magnesium and Water.

$MgSO_4 + 34\,H^2O$	22°C	199180
$MgSO_4 + 107\,H^2O$	22°C	324600

Hydrochloric Acid and Water.

$HCl + 15\,H^2O$	23°C	13626
$HCl + 500\,H^2O$	23°C	86679

233*.] MM. F. Kohlrausch and W. A. Nippoldt* have determined the resistance of mixtures of sulphuric acid and water. They used alternating magneto-electric currents, the electromotive force of which varied from $\frac{1}{2}$ to $\frac{1}{74}$ of that of a Grove's cell, and by means of a thermoelectric copper-iron pair they reduced the electromotive force to $\frac{1}{429000}$ of that of a Grove's cell. They found that Ohm's law was applicable to this electrolyte throughout the range of these electromotive forces.

The resistance is a minimum in a mixture containing about one-third of sulphuric acid.

The resistance of electrolytes diminishes as the temperature increases. The percentage increment of conductivity for a rise of 1°C is given in the following table.

* *Pogg. Ann.* cxxxviii, p. 286, Oct. 1869.

Resistance of Mixtures of Sulphuric Acid and Water at 22°C in terms of Mercury at 0°C. MM. Kohlrausch and Nippoldt.

Specific gravity at 18°5	Percentage of H_2SO_4	Resistance at 22°C (Hg=1)	Percentage increment of conductivity for 1°C.
0·9985	0·0	746300	0·47
1·00	0·2	465100	0·47
1·0504	8·3	34530	0·653
1·0989	14·2	18946	0·646
1·1431	20·2	14990	0·799
1·2045	28·0	13133	1·317
1·2631	35·2	13132	1·259
1·3163	41·5	14286	1·410
1·3547	46·0	15762	1·674
1·3994	50·4	17726	1·582
1·4482	55·2	20796	1·417
1·5026	60·3	25574	1·794

On the Electrical Resistance of Dielectrics.

234*.] A great number of determinations of the resistance of gutta-percha, and other materials used as insulating media, in the manufacture of telegraphic cables, have been made in order to ascertain the value of these materials as insulators.

The tests are generally applied to the material after it has been used to cover the conducting wire, the wire being used as one electrode, and the water of a tank, in which the cable is plunged, as the other. Thus the current is made to pass through a cylindrical coating of the insulator of great area and small thickness.

It is found that when the electromotive force begins to act, the current, as indicated by the galvanometer, is by no means constant. The first effect is of course a transient current of considerable intensity, the total quantity of electricity being that required to charge the surfaces of the insulator with the superficial distribution of electricity corresponding to the electromotive force. This first current therefore is a measure not of the conductivity, but of the capacity of the insulating layer.

But even after this current has been allowed to subside the residual current is not constant, and does not indicate the true conductivity of the substance. It is found that the current continues to decrease for at least half an hour, so that a determination of the resistance deduced from the current will give a greater value if a certain time is allowed to elapse than if taken immediately after applying the battery.

Thus, with Hooper's insulating material the apparent resistance at the end of ten minutes was four times, and at the end of nineteen hours twenty-three times that observed at the end of one minute. When the direction of the electromotive force is reversed, the resistance falls as low or lower than at first and then gradually rises.

These phenomena seem to be due to a condition of the gutta-percha, which, for want of a better name, we may call polarization, and which we may compare on the one hand with that of a series of Leyden jars charged by cascade, and, on the other, with Ritter's secondary pile.

If a number of Leyden jars of great capacity are connected in series by means of conductors of great resistance (such as wet cotton threads in the experiments of M. Gaugain), then an electromotive force acting on the series will produce a current, as indicated by a galvanometer, which will gradually diminish till the jars are fully charged.

The apparent resistance of such a series will increase, and if the dielectric of the jars is a perfect insulator it will increase without limit. If the electromotive force be removed and connexion made between the ends of the series, a reverse current will be observed, the total quantity of which, in the case of perfect insulation, will be the same as that of the direct current. Similar effects are observed in the case of the secondary pile, with the difference that the final insulation is not so good, and that the capacity per unit of surface is immensely greater.

In the case of the cable covered with gutta-percha, &c., it is found that after applying the battery for half an hour, and then connecting the wire with the external electrode, a reverse current takes place, which goes on for some time, and gradually reduces the system to its original state.

These phenomena are of the same kind with those indicated by the 'residual discharge' of the Leyden jar, except that the amount of the polarization is much greater in gutta-percha, &c. than in glass.

This state of polarization seems to be a directed property of the material, which requires for its production not only electromotive force, but the passage, by displacement or otherwise, of a considerable quantity of electricity, and this passage requires a considerable time. When the polarized state has been set up, there is an internal electromotive force acting in the substance in the reverse direction, which will continue till it has either produced a reversed current equal in total quantity to the first, or till the state of polarization has quietly subsided by means of true conduction through the substance.

The whole theory of what has been called residual discharge, absorption of electricity, electrification, or polarization, deserves a careful investigation, and will probably lead to important discoveries relating to the internal structure of bodies.

235*.] The resistance of the greater number of dielectrics diminishes as the temperature rises.

Thus the resistance of gutta-percha is about twenty times as great at 0°C as at 24°C. Messrs. Bright and Clark have found that the following formula gives results agreeing with their experiments. If r is the resistance of gutta-percha at temperature T centigrade, then the resistance at temperature $T+t$ will be

$$R = r \times 0.8878^t,$$

the number varies between 0.8878 and 0.9.

Mr. Hockin has verified the curious fact that it is not until some hours after the gutta-percha has taken its temperature that the resistance reaches its corresponding value.

The effect of temperature on the resistance of india-rubber is not so great as on that of gutta-percha.

The resistance of gutta-percha increases considerably on the application of pressure.

The resistance, in Ohms, of a cubic metre of various specimens of gutta-percha used in different cables is as follows*.

Name of Cable.	
Red Sea	$.267 \times 10^{12}$ to $.362 \times 10^{12}$
Malta-Alexandria	1.23×10^{12}
Persian Gulf	1.80×10^{12}
Second Atlantic	3.42×10^{12}
Hooper's Persian Gulf Core	74.7×10^{12}
Gutta-percha at 24°C	3.53×10^{12}

236*.] The following table, calculated from the experiments of M. Buff †, shews the resistance of a cubic metre of glass in Ohms at different temperatures.

* Jenkin's *Cantor Lectures.*

† [*Annalen der Chemie und Pharmacie,* bd. xc. 257 (1854).]

Temperature.	Resistance.
200°C	227000
250°	13900
300°	1480
350°	1035
400°	735

237*.] Mr. C. F. Varley* has recently investigated the conditions of the current through rarefied gases, and finds that the electromotive force E is equal to a constant E_0 together with a part depending on the current according to Ohm's Law, thus

$$E = E_0 + RC.$$

For instance, the electromotive force required to cause the current to begin in a certain tube was that of 323 Daniell's cells, but an electromotive force of 304 cells was just sufficient to maintain the current. The intensity of the current, as measured by the galvanometer, was proportional to the number of cells above 304. Thus for 305 cells the deflexion was 2, for 306 it was 4, for 307 it was 6, and so on up to 380, or $304+76$ for which the deflexion was 150, or 76×1.97.

From these experiments it appears that there is a kind of polarization of the electrodes, the electromotive force of which is equal to that of 304 Daniell's cells, and that up to this electromotive force the battery is occupied in establishing this state of polarization. When the maximum polarization is established, the excess of electromotive force above that of 304 cells is devoted to maintaining the current according to Ohm's Law.

The law of the current in a rarefied gas is therefore very similar to the law of the current through an electrolyte in which we have to take account of the polarization of the electrodes.

In connexion with this subject we should study Thomson's results†, in which the electromotive force required to produce a spark in air was found to be proportional not to the distance, but to the distance together with a constant quantity. The electromotive force corresponding to this constant quantity may be regarded as the intensity of polarization of the electrodes.

238*.] MM. Wiedemann and Rühlmann have recently‡ investigated the passage of electricity through gases. The electric current was produced by Holtz's machine, and the discharge took place

* *Proc. R. S.*, Jan. 12, 1871.

† [*Proc. R. S.*, 1860, or Reprint, chap. xix.]

‡ *Berichte der Königl. Sächs. Gesellschaft*, Oct. 20, 1871.

between spherical electrodes within a metallic vessel containing rarefied gas. The discharge was in general discontinuous, and the interval of time between successive discharges was measured by means of a mirror revolving along with the axis of Holtz's machine The images of the series of discharges were observed by means of a heliometer with a divided object-glass, which was adjusted till one image of each discharge coincided with the other image of the next discharge. By this method very consistent results were obtained. It was found that the quantity of electricity in each discharge is independent of the strength of the current and of the material of the electrodes, and that it depends on the nature and density of the gas, and on the distance and form of the electrodes.

These researches confirm the statement of Faraday* that the electric tension (see Art. 46) required to cause a disruptive discharge to begin at the electrified surface of a conductor is a little less when the electrification is negative than when it is positive, but that when a discharge does take place, much more electricity passes at each discharge when it begins at a positive surface. They also tend to support the hypothesis, that the stratum of gas condensed on the surface of the electrode plays an important part in the phenomenon, and they indicate that this condensation is greatest at the positive electrode.

Note on Wheatstone's Bridge.

[The following method of determining the current in the Galvanometer of Wheatstone's Bridge was given by Professor Maxwell in his last course of lectures, and is a good illustration of the method of treating a system of linear conductors. It has been communicated to the present editor by Professor J. A. Fleming of University College, Nottingham. The method simply assumes Ohm's Law for each conductor, and that the whole electromotive force around a linear circuit is the sum of the electromotive forces in the several conductors forming the circuits, and therefore equal to the sum of the products of the resistance of each conductor and the current flowing in it, the currents being taken in cyclic order.

Let P, Q, S, R, G and B (Fig. 53) denote the resistances in the several conductors forming the bridge, and let them be arranged as indicated in the figure. Now the six conductors may be considered

* *Exp. Res.*, 1501.

as forming three independent circuits viz.:—PGQ, RSG, and QSB. Let $x+y$, y and z denote the currents in these circuits respectively, each current being considered as flowing in the directions indicated by the arrows. Then the actual current in Q is $z-x-y$, that in S is $z-y$ and that in G, is x, and the electromotive force between

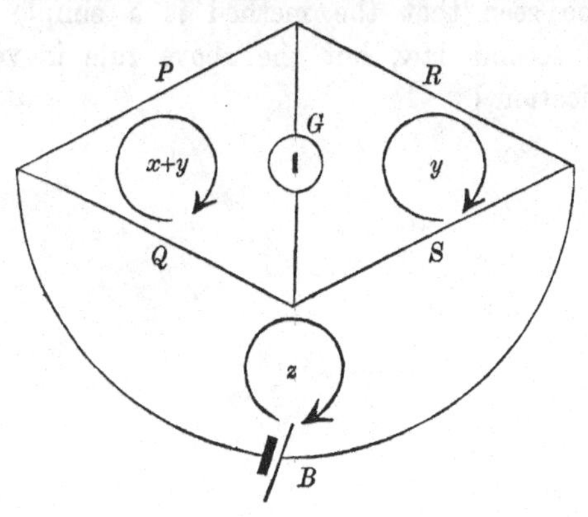

Fig. 53

the ends of Q is $Q(z-y-x)$ and so on for the other conductors. Of the three circuits specified above the E. M. F. in the first two is zero while that in the third is E, the electromotive force of the battery. Hence, applying Ohm's Law to each circuit in order we have

$$\left.\begin{array}{lll}(P+G+Q)\overline{x+y}-Gy-Qz & & =0\\ (R+S+G)y & -Sz-G\overline{x+y} & =0\\ (Q+S+B)z & -Sy-Q\overline{x+y} & =E\end{array}\right\} \quad \ldots\ldots\ldots \text{(I)}$$

or

$$\left.\begin{array}{lll}(P+G+Q)x+(P+Q)y-Qz=0\\ -Gx \quad +(R+S)y-Sz=0\\ -Qx \quad -(S+Q)y+(Q+S+B)z=E\end{array}\right\} \quad \ldots \text{(II)}$$

Solving for x we obtain

$$x=\frac{E\begin{vmatrix}\overline{P+Q}, & -Q\\ \overline{R+S}, & -S\end{vmatrix}}{\Delta}$$

$$=\frac{E(QR-PS)}{\Delta},$$

where Δ is the determinant of the system of equations (II).

The condition for no current in the galvanometer is $x=0$, or

$$QR-PS=0, \text{ or } \frac{P}{Q}=\frac{R}{S}.$$

To obtain the current equations, (I), the rule is—

'Multiply each cycle sign (i.e. current) by the sum of all the resistances which bound that cycle, and subtract from it the sign of each neighbouring cycle multiplied by the resistance separating the cycles, and equate the result to the E. M. F. in the cycle.'

It will be seen that the method is a simple application of Kirchhoff's second law, but the above rule is very convenient in its application.]

THE END.

PLATES.

PLATE I.

Art. 93.

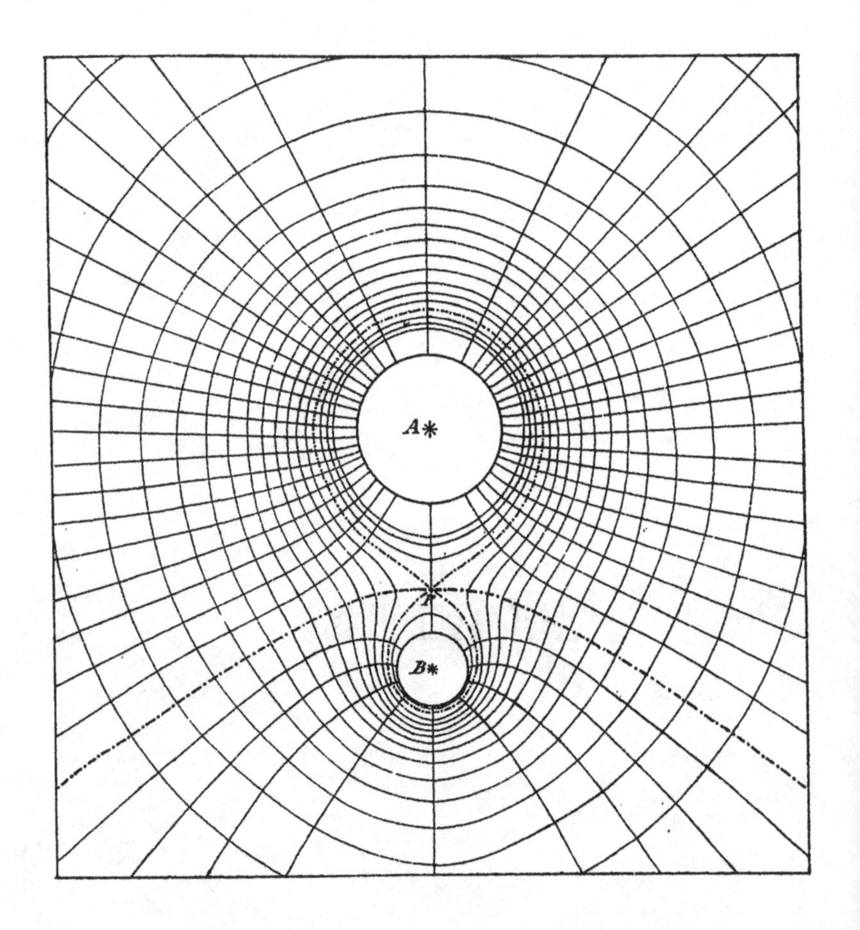

Lines of Force and Equipotential Surfaces.

A = 20 *B = 5* *P, Point of Equilibrium.* $AP = \frac{2}{3} AB$.

For the Delegates of the Clarendon Press.

PLATE II.
Art 94.

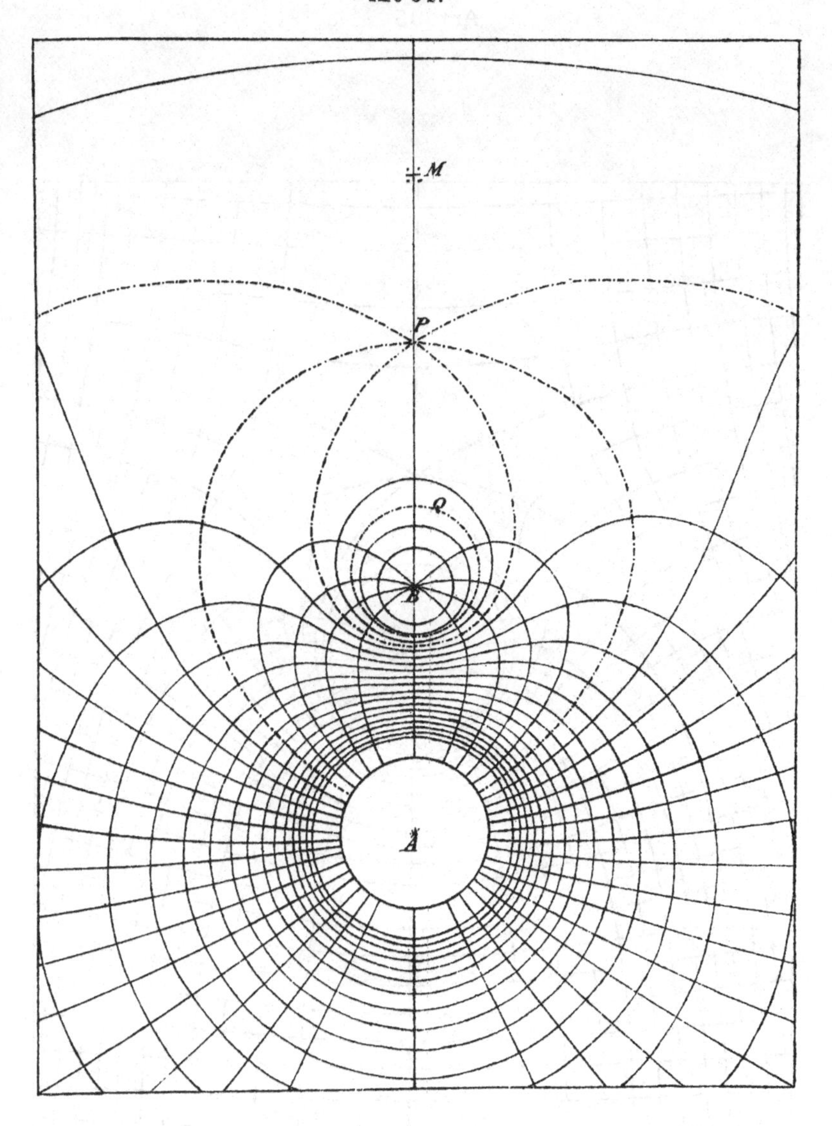

Lines of Force and Equipotential Surfaces.

A = 20   B = -5   P, Point of Equilibrium   AP = 2AB

Q, Spherical surface of Zero potential

M, Point of Maximum Force along the axis.

The dotted line is the Line of Force Ψ = 0.1̇ thus.................

For the Delegates of the Clarendon Press.

PLATE III.

Art. 95.

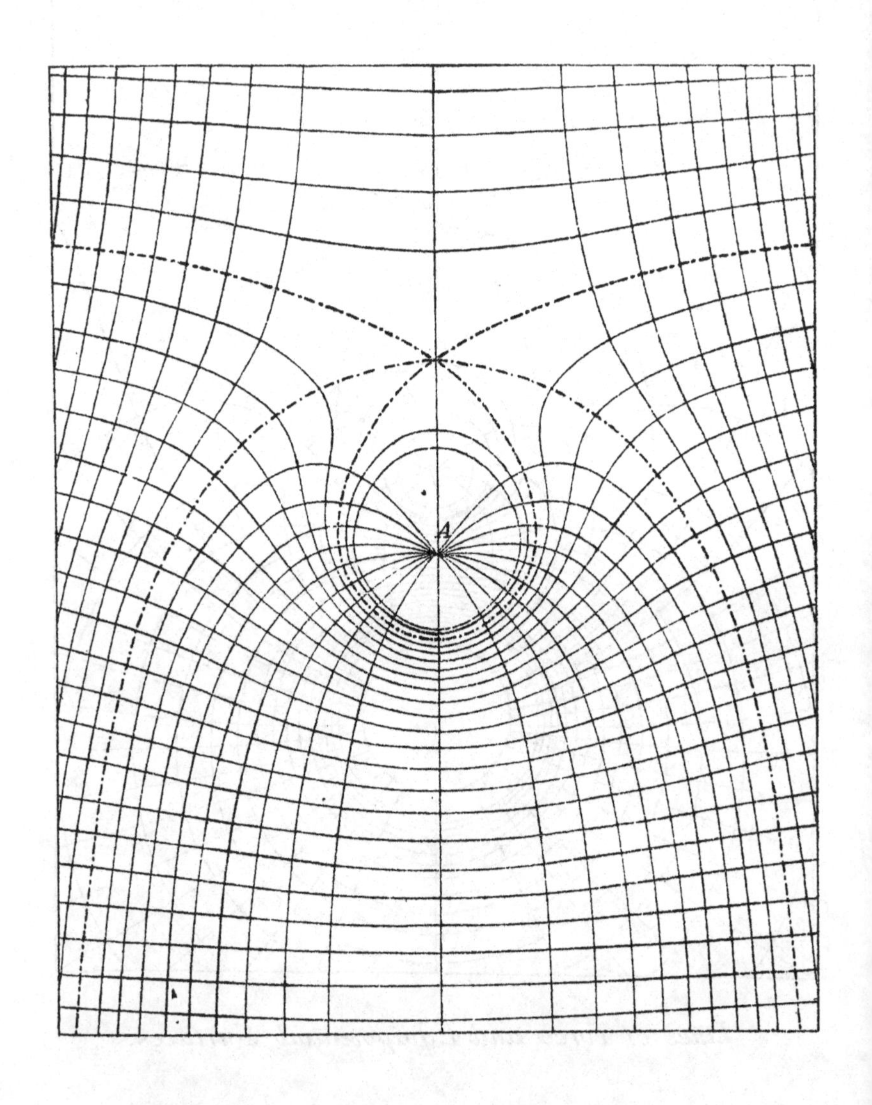

Lines of Force and Equipotential Surfaces.

A = 10

For the Delegates of the Clarendon Press.

PLATE IV.

Art. 96.

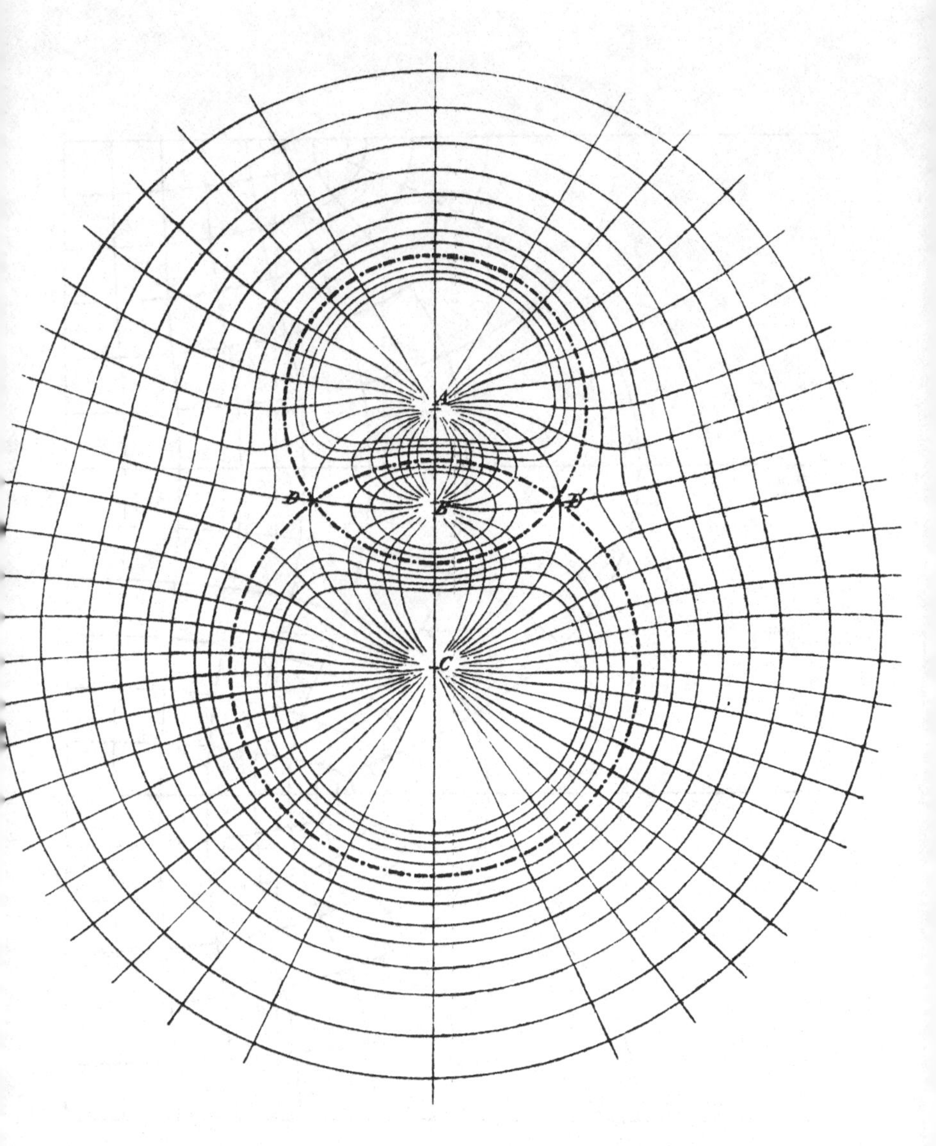

Lines of Force and Equipotential Surfaces.

A = 15 *B = −12* *C = 20.*

For the Delegates of the Clarendon Press.

Plate V.

Art 193.

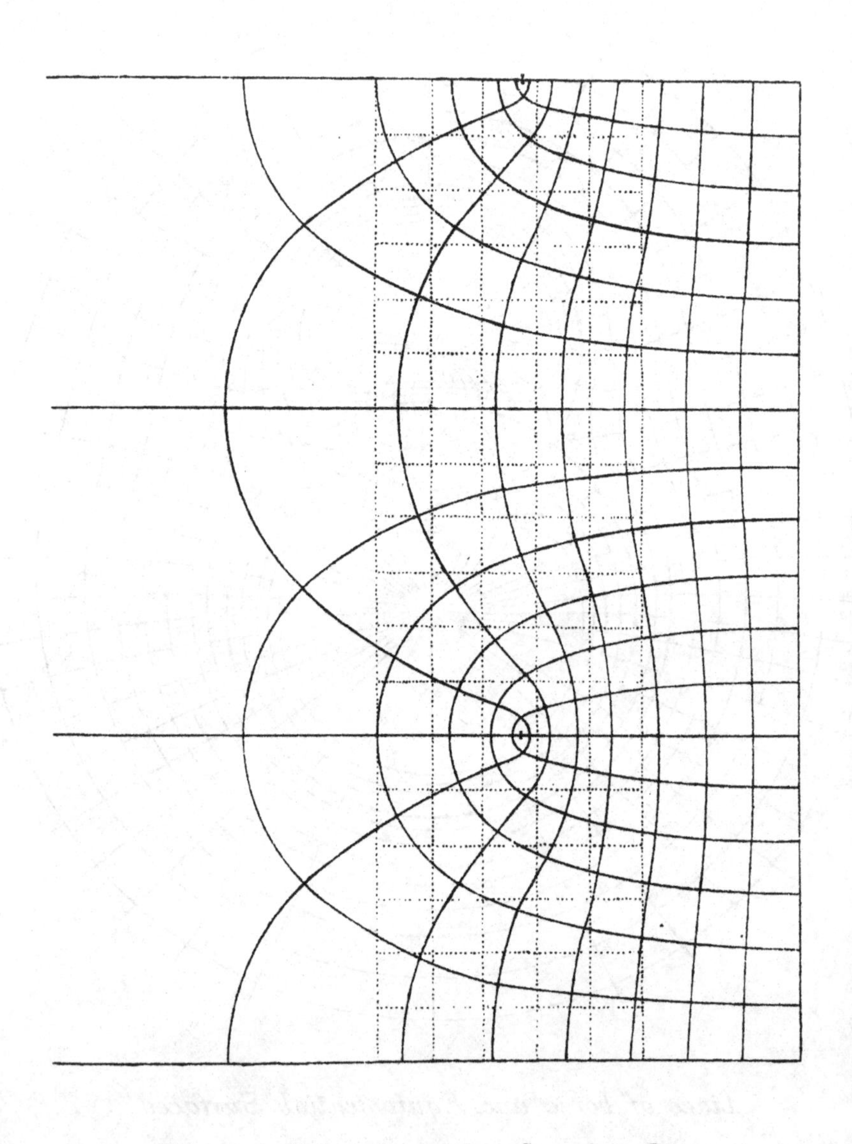

Lines of Force near the edge of a Plate.

For the Delegates of the Clarendon Press.

PLATE VI

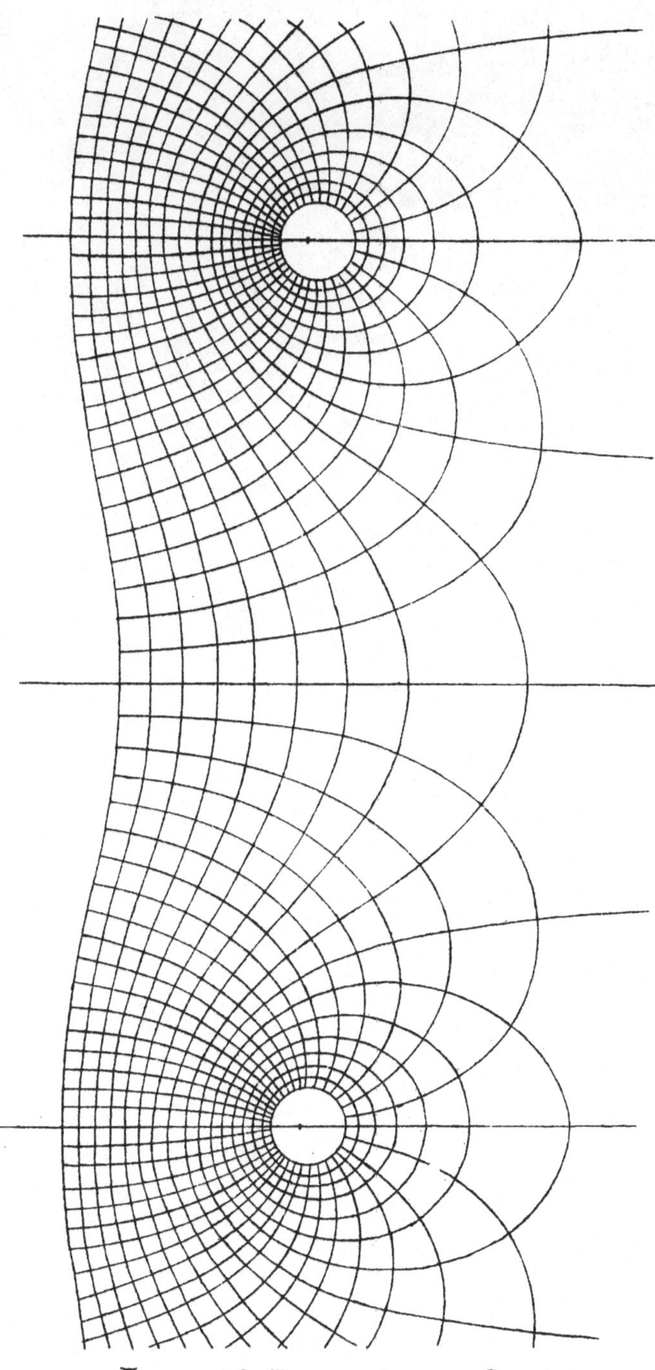

Lines of Force near a Grating

For the Delegates of the Clarendon Press.

Printed in the United States
By Bookmasters